CRITICAL PATH METHODS
IN BUILDING CONSTRUCTION

CRITICAL PATH

PRENTICE-HALL, INC., Englewood Cliffs, N.J.

METHODS
in
Building Construction

BEN BENSON

Professor and Chairman, Department of Home Building
Trinity University, San Antonio, Texas

Critical Path Methods
in Building Construction
by
Ben Benson

Current printing (last digit):
10 9 8 7 6 5 4 3 2 1

13-194001-5

Library of Congress Catalog Card Number 72-107601

Printed in the United States of America

Prentice-Hall International, Inc., *London*
Prentice-Hall of Australia, Pty. Ltd., *Sydney*
Prentice-Hall of Canada, Ltd., *Toronto*
Prentice-Hall of India Private Ltd., *New Delhi*
Prentice-Hall of Japan, Inc., *Tokyo*

PREFACE

The Critical Path Method of project scheduling is the most valuable tool ever given to management. Its potentialities were (and still are) so great that more thought, more research, and more publications were devoted to it than to any other aspect of the general construction industry.

Large industries, large construction firms, and the Department of Defense adopted it. Because the projects for which it was used were large, the firms using it were large and the computer industry soon programmed their computers for CPM. But construction terminology, construction practices, and construction men are part of a unique discipline. Construction superintendents could not grasp the new techniques, let alone computer language, and so, in a field where CPM could have been most beneficial, it was not used. There was a breakdown in liaison between the computer and the construction superintendent.

This book is written with a twofold purpose. It is primarily written for the college level student of construction to give him a working knowledge of construction project scheduling and control. It is also written to give the construction superintendent insight into a more effective tool for executing his project more economically.

The author is indebted to the late William Matera for his perspicuous analyses of several CPM schedules prepared for him and implemented by his superintendents during construction. The author wishes to thank Mr. Robert Browning of the Browning Construction Company for his valuable comments concerning a computer-oriented approach to CPM and Professor John W. Fondahl of Stanford University for permission to use material from his Technical Report No. 9, "A Non-Computer Approach to the Critical Path Method for the Construction Industry."

Ben Benson

San Antonio, Texas

CONTENTS

CRITICAL PATH METHODS
IN BUILDING CONSTRUCTION

CHAPTER 1

INTRODUCTION

The process of construction has incorporated planning since the building of the first, simplest structures. No progress or change would have occurred if men had not sought efficiency—planning in order to accomplish tasks with a minimum of effort.

Yet all the technological and social advances in construction's history—improved tools and working conditions, mechanized production methods and power-driven tools—were directed toward the production of goods; the worker was the chief beneficiary. Until the development of the critical path method, management continued the same methods throughout the centuries. Yet management's goal of project control, the ability to meet contract specifications, may depend upon a reliable method of planning a flexible, over-all schedule.

The successful resolution of a crisis that threatens to delay construction requires an immediate plan of action. At the onset of a problem the plan must be put forth immediately by the foreman, who should be chosen for leadership abilities. However, his intuition and personal experience are not proper bases for formulating the plan; it must be thought out carefully step by step, illustrated graphically, and seen and understood by all involved before the job is begun.

Planning was required for the first known building construction of significance, the pyramids of Egypt; stones had to be quarried, cut to predetermined size, transported to the construction site, and installed. Ramps were built for scaffolding and hoisting and were removed at the pyramids' completion. The "superintendent" had not only to schedule all the proposed construction but also to pass on plans and schedules to his successor.

Perhaps the first construction schedules were scratches made in the earth or on rocks. Later the schedules were graphic. More recently graphics were combined with cost controls. Then cost controls were combined with quantity controls; quantity controls were later combined with percentage of completion. These were followed by bar charts; bar charts were combined with lazy S curves and finally with CPM. It is the intent of this book to demonstrate that the best method yet of job scheduling and control is a composite of CPM and bar chart.

Each of these methods of progress scheduling and job control was an improvement on its preceding method, and therefore had some advantages over it. And yet each method failed in some way when put into practice. The color-graph (a graphic method of control) was semipermanent and could show very well the progress of the work, but it could not show whether the work was progressing ahead of or behind schedule. Neither was it able to demonstrate a gain or loss over estimated costs.

Cost control provided the answer to unit costs as well as to total cost of each activity, but it could not correlate these costs with respect to quantity of work of each activity. The building contractor still could not foretell a profit or loss over the entire project. The best that cost control can do is to provide the contractor with a system of unit costs upon which to base his bid for similar buildings in the future.

Quantitative records needed to be kept and scanned in perspective with cost control.

In order to control a building project, or rather to compare progress with total estimate, a system of percentages of completions was devised. This will be described later. The percentage of completion, although an improvement on previous methods, was found to be lacking in a most important area. Many buildings whose construction was executed efficiently were found to be 110% or more completed and not yet ready to be released to the owner. Percentage of completion methods were not abandoned entirely but generally gave way to the more popular bar chart.

The bar chart was a graphic presentation in the form of bars for the major operations of a building construction project. The bar chart showed the anticipated starting and completion dates of each major operation. But the interrelationship of operations was maintained strictly in the mind of the project superintendent. Current progress reporting might show that although some operations were being performed ahead of schedule the project as a whole was felt to be behind schedule. There really was no better way than "feel" to control the project. Then, too, it was found as often as not that those operations being completed physically ahead of schedule were costing the contractor more money than he had estimated originally. In some not too rare cases, an entire building project was completed well ahead of schedule at a loss to the contractor, and that loss could not be pinpointed.

Construction superintendents have always disagreed over which is the more difficult part of a job: getting out of the ground (foundation work) or completing a job. One fact is clear: both parts are considerably slower than the big middle of a job. So, by superimposing a percentage of cost completion curve over a bar chart, it was found that the curve resembled a lazy S. The advantages of the lazy S curve far outweigh its disadvantages insofar as progress reporting is concerned, even in planning operations—but only in a vague way because the lazy S method does not show the interrelationship of individual operations and the relationship of operations to the whole project.

Then came the critical path method of scheduling building construction. The CPM is not the perfect control the building contractor would like. It cannot reflect directly unit costs of operations or total cost of the project to date. But it does provide a tangible means of comparing operations so that a positive determination can be made of the most economical operation when a change in schedule is indicated.

A small warehouse, one-story, 55'-0" x 150'-0", will be used for examples and demonstrations throughout this book. A typical wall section of this building is shown in Figure 1-1 and a shortened quantity survey and cost estimate for it are presented in Figure 1-2.

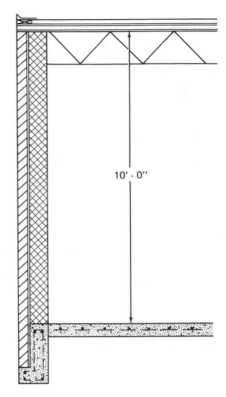

10' - 0"

FIGURE 1-1 Typical wall section of a small warehouse.

GENERAL BUILDERS

Job: A Small Warehouse   Cost Estimate

Item of Work	Quantity	Unit	Price	Total	Prorated
Job layout	450	L.F.	0.20	90	108
Excavation	125	C.Y.	4.00	500	600
Formwork	4,200	S.F.	1.00	4,200	5,040
Reinforcing	8,000	Lb.	0.11	880	1,056
Concrete	218	C.Y.	16.00	3,488	4,186
Cement finishing	8,250	S.F.	0.07	578	693
Masonry	10,020	S.F.	1.14	11,423	13,708
Waterproofing	4,920	S.F.	0.15	738	886
O.W. steel joists	12	Ton	350.00	4,200	5,040
Roofing, decking & sheet metal	83	SQ.	66.00	5,478	6,573
Metal doors & windows	28	EA.	60.00	1,680	2,016
Carpentry	800	B.F.	0.20	160	192
Painting	175	SQ.	10.00	1,750	2,100
Mechanical	Lump	Sum		6,000	7,200
Electrical	Lump	Sum		3,500	4,200
				$44,665	
Overhead & profit	20	%		8,933	
TOTAL CONTRACT PRICE					$53,598

FIGURE 1-2

CHAPTER 2

METHODS
OF
PROJECT CONTROL

Graphic Methods

There are several graphic methods of control, each useful in its particular way depending on the objective. One of the earliest known methods is referred to as color-graph.

The procedure for color-graph control of a building project is as follows. Blue line prints of an outline of the entire building are displayed in both the home and field offices. Several copies in blue line are kept at the field office for reporting periodic progress in colors to the home office. On large or multistory constructions these reporting prints need to show only one typical story, all in outline form. For a housing project the reporting print need outline one house only. The master plan, of course, shows all the stories of the high-rise and numbers each floor; or it shows, in plan, all of the houses and numbers them.

Color pencils or crayons are used to indicate progress; for example, black for excavation and fill, yellow for formwork, blue for steel, green for concrete. At the end of a reporting period, the job superintendent colors his displayed master plan. His assistant or timekeeper then indicates by color the progress on individual reporting prints, identifying each individual print by house or story number and reporting date. The reporting prints are forwarded to the home office where the information contained in them is transferred to the home office master plan. The builder can then see the work progressing.

At the same time the timekeeper reports his payrolls in a fashion that shows how much money was spent for each "color" of work, as well as weekly and total to date payrolls. The builder then can add his materials bills to his payrolls and compare the cost of the work to date with the amount of work to date as shown on the master plan.

Home builders still use this technique, it is modified by the use of colored pins that show foundations, house completions, and sales.

Color-graphs are adequate for an over-all but relatively loose control of projects. It is not in sufficient detail for a project such as a small warehouse. A percentage of completion method would be better.

Percentage of Completion

Simply stated, a percentage of completion method of job progress control is based upon a comparison of money spent against estimated cost for each major item of work as well as for the total project. This comparison is made by use of a form or forms that are standardized within the company. The form is filled in initially by the estimator (see Figure 2-1) who then gives it along with a supply of blank forms to the project superintendent. The superintendent or his job office assistant will use this form to report progress periodically to the home office. The best period is weekly; however, monthly reporting may suffice for large projects.

The mechanics for reporting progress based on a percentage of completion in terms of cost or money spent are demonstrated in Figures 2-2, 2-3, and 2-4.

For example, at the end of the first week, work items 1 and 2 are completed and it is estimated that item 3 is 33% completed. It makes no difference here that 135 C. Y. of excavation may have been required to complete the item, which was estimated to have required only 125 C. Y.; it is nevertheless 100% completed.

GENERAL BUILDERS

Job : Small Warehouse   Progress Report No. _____ for Week Ending _____

No.	Item of Work	Contract Estimate					This Period			Total to Date			Remarks
		Quantity	Unit	Price	Amount	% of Total	% of Work Completed	Amount Spent	% of Total	% of Work Completed	Amount Spent	% of Total	
1.	Job layout & batterboards	450	L.F.	0.20	90	0.2							
2.	Excavation	125	C.Y.	4.00	500	1.1							
3.	Formwork	4,200	S.F.	1.00	4,200	9.4							
4.	Reinforcing	8,000	Lb.	0.11	880	2.0							
5.	Concrete	218	C.Y.	16.00	3,488	7.8							
6.	Cement finishing	8,250	S.F.	0.07	578	1.3							
7.	Masonry	10,020	S.F.	1.14	11,423	25.5							
8.	Waterproofing	4,920	S.F.	0.15	738	1.7							
9.	O.W. steel joists	12	Ton	350.00	4,200	9.4							
10.	Roof, decking & sheet metal	83	SQ.	66.00	5,478	12.3							
11.	Metal doors & windows	28	EA.	60.00	1,680	3.8							
12.	Carpentry	800	B.F.	0.20	160	0.4							
13.	Painting	175	SQ.	10.00	1,750	3.9							
14.	Mechanical		Lump Sum		6,000	13.4							
15.	Electrical		Lump Sum		3,500	7.8							
	TOTALS				44,665	100.0							

FIGURE 2-1

GENERAL BUILDERS

Job : Small Warehouse   Progress Report No. _____ for Week Ending _____

No.	Item of Work	Contract Estimate					This Period			Total to Date			Remarks
		Quantity	Unit	Price	Amount	% of Total	% of Work Completed	Amount Spent	% of Total	% of Work Completed	Amount Spent	% of Total	
1.	Job layout & batterboards	450	L.F.	0.20	90	0.2	100	95	0.2	100	95	0.2	
2.	Excavation	125	C.Y.	4.00	500	1.1	100	567	1.3	100	567	1.3	135 C.Y. required to complete
3.	Formwork	4,200	S.F.	1.00	4,200	9.4	33	1,400	3.1	33	1,400	3.1	
4.	Reinforcing	8,000	Lb.	0.11	880	2.0							
5.	Concrete	218	C.Y.	16.00	3,488	7.8							
6.	Cement finishing	8,250	S.F.	0.07	578	1.3							
7.	Masonry	10,020	S.F.	1.14	11,423	25.5							
8.	Waterproofing	4,920	S.F.	0.15	738	1.7							
9.	O.W. steel joists	12	Ton	350.00	4,200	9.4							
10.	Roof, decking & sheet metal	83	SQ.	66.00	5,478	12.3							
11.	Metal doors & windows	28	EA.	60.00	1,680	3.8							
12.	Carpentry	800	B.F.	0.20	160	0.4							
13.	Painting	175	SQ.	10.00	1,750	3.9							
14.	Mechanical	Lump Sum			6,000	13.4							
15.	Electrical	Lump Sum			3,500	7.8							
	TOTALS				44,665	100.0	4.4	2,062	4.6	4.4	2,062	4.6	

FIGURE 2-2

Job : Small Warehouse GENERAL BUILDERS Progress Report No. _____ for Week Ending _____

No.	Item of Work	Contract Estimate					This Period			Total to Date			Remarks
		Quantity	Unit	Price	Amount	% of Total	% of Work Completed	Amount Spent	% of Total	% of Work Completed	Amount Spent	% of Total	
1.	Job layout & batterboards	450	L.F.	0.20	90	0.2				100	95	0.2	
2.	Excavation	125	C.Y.	4.00	500	1.1				100	567	1.3	
3.	Formwork	4,200	S.F.	1.00	4,200	9.4	42	1,675	3.8	75	3,075	6.9	
4.	Reinforcing	8,000	Lb.	0.11	880	2.0							
5.	Concrete	218	C.Y.	16.00	3,488	7.8							
6.	Cement finishing	8,250	S.F.	0.07	578	1.3							
7.	Masonry	10,020	S.F.	1.14	11,423	25.5							
8.	Waterproofing	4,920	S.F.	0.15	738	1.7							
9.	O.W. steel joists	12	Ton	350.00	4,200	9.4							
10.	Roof, decking & sheet metal	83	SQ.	66.00	5,478	12.3							
11.	Metal doors & windows	28	EA.	60.00	1,680	3.8							
12.	Carpentry	800	B.F.	0.20	160	0.4							
13.	Painting	175	SQ.	10.00	1,750	3.9							
14.	Mechanical	Lump Sum			6,000	13.4							
15.	Electrical	Lump Sum			3,500	7.8							
	TOTALS				44,665	100.0	3.9	1,675	3.8	8.4	3,737	8.4	

FIGURE 2-3

GENERAL BUILDERS

Job : Small Warehouse Progress Report No. _____ for Week Ending _____

No.	Item of Work	Contract Estimate					This Period			Total to Date			Remarks
		Quantity	Unit	Price	Amount	% of Total	% of Work Completed	Amount Spent	% of Total	% of Work Completed	Amount Spent	% of Total	
1.	Job layout & batterboards	450	L.F.	0.20	90	0.2				100	95	0.2	
2.	Excavation	125	C.Y.	4.00	500	1.1				100	567	1.3	
3.	Formwork	4,200	S.F.	1.00	4,200	9.4				100	4,408	9.9	
4.	Reinforcing	8,000	Lb.	0.11	880	2.0				100	919	2.1	
5.	Concrete	218	C.Y.	16.00	3,488	7.8				100	3,706	8.3	
6.	Cement finishing	8,250	S.F.	0.07	578	1.3				100	536	1.2	
7.	Masonry	10,020	S.F.	1.14	11,423	25.5				100	11,751	26.3	
8.	Waterproofing	4,920	S.F.	0.15	738	1.7				100	740	1.7	
9.	O.W. steel joists	12	Ton	350.00	4,200	9.4				100	4,530	10.1	
10.	Roof, decking & sheet metal	83	SQ.	66.00	5,478	12.3				100	5,478	12.3	
11.	Metal doors & windows	28	EA.	60.00	1,680	3.8				67	765	1.7	
12.	Carpentry	800	B.F.	0.20	160	0.4	100	195	0.4	100	195	0.4	
13.	Painting	175	SQ.	10.00	1,750	3.9				0			
14.	Mechanical	Lump Sum			6,000	13.4	40	2,400	5.4	100	6,000	13.4	
15.	Electrical	Lump Sum			3,500	7.8	70	2,450	5.5	100	3,500	7.8	
	TOTALS				44,665	100.0	11.3	5,045	11.3	94.9	43,190	96.7	

FIGURE 2-4

To determine the total percentage of work completed, multiply the percent of total of each item of work by the percent of work completed for each item, and add the products; e.g.,

$$
\begin{aligned}
\text{item 1.} \quad & 0.2 \times 1.00 = 0.2\% \\
\text{item 2.} \quad & 1.1 \times 1.00 = 1.1\% \\
\text{item 3.} \quad & 9.4 \times 0.33 = \underline{3.1\%} \\
\text{Total \% of work completed} & = 4.4\%
\end{aligned}
$$

Comparing this 4.4% of total work completed with the 4.6% of estimate spent, one can assume that this job is not starting out very well. The remarks column reveals that there has been an overrun of estimated quantity of excavation. But the report also reveals that excavation has cost $4.20 per C.Y. against an estimate of $4.00 per C.Y. However, the item of work has been completed and nothing can be done now to pull it out of trouble.

At the end of each succeeding week the three columns headed by "Total to Date" are entered first. Then, for each item of work, the "Total to Date" of the preceding week is deducted in order to arrive at the figures for "This Period." This method avoids the possibility of carrying over an error from week to week.

In Figure 2-3, Progress Report Number 2 for the end of the second week, items 1 and 2 are completed; formwork, item 3, is now about 75% completed at a cost of $3,075, or 73% of the estimate for this item. Deducting the 33% and $1,400 of item 3 of the previous week gives 42% and $1,675 as all of the work performed and total cost during this reporting period. Following the procedure already established for the determination of percent of work completed and summing the columns of % of Total, the builder now concludes that this was a good week with 3.9% of the work performed at 3.8% of the cost and, to date, 8.4% of the work has been performed at 8.4% of the cost. It appears that, as of the end of the second week, enough money has been saved on formwork to offset the amount lost in excavation.

By the end of the tenth week, as reported in Figure 2-4, we can see the general trend. By a comparison, item by item, of the percent of total to date against the contract estimate, we now can tell which items of work have overrun and underrun. Item 5, concrete, shows a loss of $218, whereas item 6, cement finishing, shows a slight profit.

So far we have spent 96.7% of our estimate, but only 94.9% of work has been completed. A close examination of this report indicates that when the job is 100% completed, we probably will have spent some $45,322, which is 101.5 percent of the contract estimate. The amounts shown in the contract estimate for each item of work do not include overhead and profit, so the expenditure of 101.5 percent indicates that the job would not yield quite the anticipated profit.

In this method of control, when an item of work is completed, it is reported as 100 percent complete. But the quantity of work performed may have varied more or less than the estimated amount, and so we have lost control of our unit prices. It is not enough to know whether we over-estimated or underestimated our unit prices; we need to know exactly what our unit price per item is in order to estimate subsequent jobs more efficiently.

Another control of the project could have been obtained by basing the percentage of completion on the value of work performed rather than on the amount of money spent. This necessitates accurate measurements of the work performed during each reporting period. This is accomplished by a survey of the work done to date for each item and deducting from that quantity the previous total to date to arrive at the quantity of an item performed during the reporting period. The same procedure obtains for determining the value. In order for the value to reflect a percentage of the project, the quantity of work performed on each item is multiplied by the estimated price, as shown in figure 2-5.

This form could be elaborated to show actual monies spent and unit costs developed because it does record quantities of work. Examining each item, we see that some quantities overrun the estimate and some underrun, but we cannot tell whether the item of work is physically complete or not. This information could be shown under Remarks.

The foregoing methods in their way show a financial status for each reporting period and are adequate, with overhead and profit added, for use by management to request and secure approval of periodic draws. A draw is a periodic payment by the owner to the builder based on a specified value of labor and material incorporated in the job and/or delivered to the site. But, so far we have not been able to control the project with respect to time.

Job : Small Warehouse

GENERAL BUILDERS

Progress Report No. _____   for Week Ending _____

No.	Item of Work	Contract Estimate					This Period			Total to Date			Remarks
		Quantity	Unit	Price	Amount	% of Total	Quantity Completed	Value	% of Total	Quantity Completed	Value	% of Total	
1.	Job layout & batterboards	450	L.F.	0.20	90	0.2				450	90	0.2	
2.	Excavation	125	C.Y.	4.00	500	1.1				135	540	1.2	
3.	Formwork	4,200	S.F.	1.00	4,200	9.4				4,200	4,200	9.4	
4.	Reinforcing	8,000	Lb.	0.11	880	2.0				8,000	880	2.0	
5.	Concrete	218	C.Y.	16.00	3,488	7.8				229	3,664	8.2	
6.	Cement finishing	8,250	S.F.	0.07	578	1.3				8,250	578	1.3	
7.	Masonry	10,020	S.F.	1.14	11,423	25.5	3,660	4,172	9.3	6,600	7,524	16.8	
8.	Waterproofing	4,920	S.F.	0.15	738	1.7							
9.	O.W. steel joists	12	Ton	350.00	4,200	9.4							
10.	Roof, decking & sheet metal	83	SQ.	66.00	5,478	12.3							
11.	Metal doors & windows	28	EA.	60.00	1,680	3.8							
12.	Carpentry	800	B.F.	0.20	160	0.4							
13.	Painting	175	SQ.	10.00	1,750	3.9							
14.	Mechanical	Lump Sum			6,000	13.4				25%	1,500	3.4	
15.	Electrical	Lump Sum			3,500	7.8							
	TOTALS				44,665	100.0		4,172	9.3		18,976	42.5	

FIGURE 2-5

Time is money. Time may very well be the most important factor to consider. Building contracts usually have time delay damage clauses whereby a specific amount of money is deducted from the amounts due the contractor for every day the completion of the building goes beyond the specified completion date. The home builder pays interest on his construction loan. Prolonging the completion of his subdivision, which delays sales of his homes, increases the dollar amount of interest he must pay on his loans. Also, overhead expense increases and profit shrinks each day construction work is extended. And so, effective job planning and progress control are necessary.

A schedule is a proposed plan of operation showing the anticipated starting and completion dates of all the items of work constituting a project. Ideally, the schedule should reflect the most efficient use of manpower and equipment required to complete the project on time. This seldom is the case. Usually the schedule reflects the most economical use of manpower and equipment readily available to the builder that also will allow him to complete the project on time.

Properly interpreted, a schedule aids management in determining when and how much labor and equipment will be needed for specific items of work. It could aid in purchasing and ordering materials for timely deliveries.

A schedule should show the item of work and its identifying number, as well as its quantity and unit of measure. It should show its size or percent relation to the entire project. It should show the duration (time required to complete the item) as well as its starting and completion dates.

Progress control is exercised by the project superintendent in determining actual construction progress against planned progress. His objective is to see that the work proceeds economically toward timely completion of the project. To accomplish this purpose, the superintendent needs to check constantly actual work progress against planned progress. He does this by physically measuring the work done daily and periodically comparing the total of work to date with that scheduled. Should any considerable deviation appear, he would take steps to correct it. If an item of work is taking more time or less time to perform, the superintendent will take actions to correct the condition.

It already has been seen how job progress can be moderately controlled by percentages of completion. The bar chart method will be discussed next.

Bar Chart

A bar chart, sometimes called a Gantt chart after the man who first developed it early in the century, is a series of bars showing the anticipated starting time and completion dates of the several items of work that make up a project. Most building construction project contracts require the builder to submit to the architect for his approval a construction progress schedule showing the starting and completion dates of the salient items of work. This schedule shows the architect how the builder proposes to construct the project within the total time specified in the contract.

The builder drafts this schedule in the form of a bar chart, as in Figure 2-6. At the left-hand column, he lists the items of work in the order in which he finds them in his quantity survey (perhaps omitting some of the smaller ancillary items) showing the quantity and unit of each item. If the contract specifications require the builder to submit his cost breakdown for the several items of work, two additional columns may suffice for showing his unit prices and total costs after ancillary items, overhead, and profit have been distributed. Architects need to have this information as a basis for negotiating change orders to the contract, as well as for certification of progress payments.

Figure 2-6 contains a seventh column headed Time. This column need not always be used as it is redundant. The bar length is sufficient to show the duration of the pertinent items of work. The next step is to lay out to scale the project duration. This can be done in working days, calendar days, or the calendar itself. It usually is laid out in conformity with the contract time. Contracts may specify so many working days, or so many calendar days, or completion by a specific date. It is now incumbent upon the project superintendent to analyze his job, note the estimated quantities of each item of work, and show a bar to the right of each item from his anticipated starting date to completion

Job: Small Warehouse

Progress Schedule in Working Days

No.	Item of Work	Quantity	Unit	Price	Amount	Time
1.	Job layout	450	L.F.	0.24	108	1
2.	Excavation	125	C.Y.	4.80	600	3
3.	Formwork	4,200	S.F.	1.20	5,040	12
4.	Reinforcing	8,000	Lb.	0.132	1,056	2
5.	Concrete	218	C.Y.	19.20	4,186	1
6.	Cement finish	8,250	S.F.	0.084	693	1
7.	Masonry	10,020	S.F.	1.368	13,708	15
8.	Waterproofing	4,920	S.F.	0.18	886	5
9.	O.W. steel joists	12	Ton	420.00	5,040	5
10.	Deck, roof, & s/m	83	SQ.	79.19	6,573	4
11.	Metal doors & windows	28	EA.	72.00	2,016	3
12.	Carpentry	800	B.F.	0.24	192	2
13.	Painting	175	SQ.	12.00	2,100	10
14.	Mechanical	Lump	Sum		7,200	13
15.	Electrical	Lump	Sum		4,200	7
	TOTAL				53,598	

Contract Estimate

Progress Schedule columns (in Working Days): 5, 10, 15, 20, 25, 30, 35, 40, 45, 50, 55, 60, 65

FIGURE 2-6

date of each item. There is danger that the schedule may be squeezed in to show completion of the project within the contract time, when in reality the builder's resources of labor, equipment, and subcontractors are not of the capacity or bent to do so. All builders have their strengths and weaknesses, and these qualities vary from builder to builder, enabling builder A to perform one job more economically than builder B. Yet, on some other job, the tables may be turned.

Some progress schedules may show the bars as heavy solid lines, others may show them as hollow bars. Both ways have advantages. If the bar is shown solid, then, after construction gets underway, progress is indicated with another bar of a different color immediately above or below the scheduled bar. If a hollow bar is used, physical progress can be indicated by filling in the bars, as shown in Figure 2-7. Progress of individual items of work can be noted by making a comparison of the progress bar with the scheduled bar. But over-all project progress is not so readily discernible.

Bars may show some items to be on schedule, others ahead of schedule, and yet others behind schedule. The superintendent needs to analyze the entire project to see if the items behind schedule would delay the start of subsequent items and hence delay their completion. The bar chart in itself is not adequate for this.

Lazy S Curve

A way to improve the bar chart is to superimpose a lazy S curve, as in Figure 2-8. To construct a lazy S curve, assume that the cost of any item of work is uniformly distributed over its duration. This is not a bad assumption because the make-up of labor crews and use of equipment do not vary during the duration of the item of work. The length of bar, although time scaled to its duration, can be converted to its amount in dollars. With a five-day work week as one calendar week, at the end of each week add up vertically the dollar volume represented by all bars to the left of each week and plot the point to a dollar scale, vertically. The column to the extreme right in Figure 2-8 has been graduated to a dollar scale using the vertical size of the bar chart as a basis.

Job: Small Warehouse

No.	Item of Work	Quantity	Unit	Price	Amount	Time
1.	Job layout	450	L.F.	0.24	108	1
2.	Excavation	125	C.Y.	4.80	600	3
3.	Formwork	4,200	S.F.	1.20	5,040	12
4.	Reinforcing	8,000	Lb.	0.132	1,056	2
5.	Concrete	218	C.Y.	19.20	4,186	1
6.	Cement finish	8,250	S.F.	0.084	693	1
7.	Masonry	10,020	S.F.	1.368	13,708	15
8.	Waterproofing	4,920	S.F.	0.18	886	5
9.	O.W. steel joists	12	Ton	420.00	5,040	5
10.	Deck, roof, & s/m	83	SQ.	79.19	6,573	4
11.	Metal doors & windows	28	EA.	72.00	2,016	3
12.	Carpentry	800	B.F.	0.24	192	2
13.	Painting	175	SQ.	12.00	2,100	10
14.	Mechanical	Lump Sum			7,200	13
15.	Electrical	Lump Sum			4,200	7
	TOTAL				53,598	

Contract Estimate

Progress Schedule in Working Days (5, 10, 15, 20, 25, 30, 35, 40, 45, 50, 55, 60, 65)

FIGURE 2-7

Job: Small Warehouse

No.	Item of Work	Quantity	Unit	Price	Amount	Time
1.	Job layout	450	L.F.	0.24	108	1
2.	Excavation	125	C.Y.	4.80	600	3
3.	Formwork	4,200	S.F.	1.20	5,040	12
4.	Reinforcing	8,000	Lb.	0.132	1,056	2
5.	Concrete	218	C.Y.	19.20	4,186	1
6.	Cement finish	8,250	S.F.	0.084	693	1
7.	Masonry	10,020	S.F.	1.368	13,708	15
8.	Waterproofing	4,920	S.F.	0.18	886	5
9.	O.W. steel joists	12	Ton	420.00	5,040	5
10.	Deck, roof, & s/m	83	SQ.	79.19	6,573	4
11.	Metal doors & windows	28	EA.	72.00	2,016	3
12.	Carpentry	800	B.F.	0.24	192	2
13.	Painting	175	SQ.	12.00	2,100	10
14.	Mechanical	Lump	Sum		7,200	13
15.	Electrical	Lump	Sum		4,200	7
	TOTAL				53,598	

Contract Estimate

Progress Schedule in Working Days

FIGURE 2-8

For example, at the end of the first week it is seen that all of item 1, $108, plus all of item 2, $600, and 4/12 of item 3, or 4/12 of $5,040, or $1,680, add up to $2,388. The first point has been plotted at 0,0; this second point is plotted horizontally at week 1 and vertically at $2,388. Similarly, by the end of the second week, to the previous total of $2,388 another 5/12 of $5,040 is added giving the third point—horizontally at week 2 and vertically at $4,488. And so on until a point has been plotted for each week of the project. The points now are joined by straight lines. The curve, so developed, resembles a lazy S. The irregularities in this curve now are smoothed out with a French curve without obliterating the original straight line work.

It is now worthwhile to make a comparison of the two curves. Because the construction of a building begins slowly, gains momentum, progresses steadily, and then ends slowly, the curve developed from the bar chart should reflect this progression. If there are any appreciable deviations of the plotted curve from the smoothed curve, the chances are that a serious mistake has been made, either in estimating the cost of the project or in planning its execution. If the plotted curve is below the smooth curve, either a large item of work has been left out of the estimate or the work has been planned to progress too slowly. If the plotted curve is above the smooth curve, then one or more items of work have been planned to progress too rapidly. Either way, the portent spells trouble, which should be eliminated before construction is begun.

After construction has been started, control is maintained by plotting the construction curve in the same manner as and over the planned curve, periodically, and comparing the two. If the construction curve lies above the planned curve and the construction bars are on schedule, then the work is costing more money than estimated. If the construction bars are shorter than the planned bars, the job is progressing ahead of schedule. If the construction bars are longer than the planned bars, then the job is losing money. If the construction curve falls below the planned curve, then the project either is falling behind schedule or is earning greater profit than planned, again depending upon the relation of construction to planned bars.

When a project is in trouble and behind schedule, too many superintendents tend to stomp on their hats, or to take off in all directions like a covey of quail. They will speed up all items of work by going into overtime, enlarging the working force, and adding more equipment—much of which adds to the cost and does not contribute to reduction of the project duration. The critical path method is the only one which can be used in selecting the one or more specific items of work which can be speeded up and thus reduce the project duration.

CHAPTER 3

LOGIC OF
NETWORK DIAGRAMMING

To use the critical path method effectively, one must have knowledge of both **CPM** and tectonics. In building construction, **CPM** without tectonics goes awry. There are two basic presentations of CPM, the arrow diagram and the circle notation. The circle notation is easier to work with; the arrow diagram is easier to read and to understand. The logic in diagramming either is the same. So the arrow network will be used for explanation of the logic.

The objective of a network is to build the project on paper, step by step, exactly as it would be planned for construction. The starting point for developing a network is a competent quantity survey. The estimator has analyzed already the construction drawings or blueprints in order to prepare the quantity survey and cost estimate, so he is the person to develop the network. He will do this without regard to time, that is, without regard to the duration of the project or any of its items of work. He will use an arrow having circles at the head and tail:

At this time the length of arrow is of no consequence, providing that it is long enough for a written description of the item of work, which from now on will be called an activity. It helps to understand the process at this time to make an arbitrary distinction between <u>item of work</u> and <u>activity</u>. An item of work refers to the listings in a quantity survey; activity refers to an element of work between nodes of an arrow diagram. Later the distinction will be broadened to include the operation of the circle notation network. Every construction project must have a beginning, even if it is nothing more than a notice to proceed with the work, and an ending, if nothing more than a punch list, or cleaning up and moving out or "selling" the job. Between the start and end must fall everything required to construct in accordance with the plans and specifications. It often has been said that a job is losing money any time that nothing is being done. Stated positively, there always is some work that can be done on a construction job. This truth then leads to the three fundamentals, or questions, that need to be answered for every activity:

1. What activity must immediately precede the start of this activity?
2. What activity must immediately follow this activity?
3. What activity or activities may be performed simultaneously with this activity?

The critical path method of project scheduling must adhere strictly to this format in order to be meaningful and effective.

In order to draft and read an arrow diagram in the light of the three fundamentals, certain ground rules have been developed; for example:

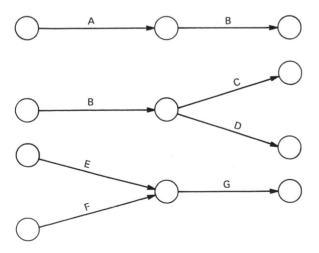

Activity B cannot start until activity A is completed.

Activity B is immediately followed by activities C and D, which may be performed concurrently.

Activity G cannot start until both activities E and F have been completed.

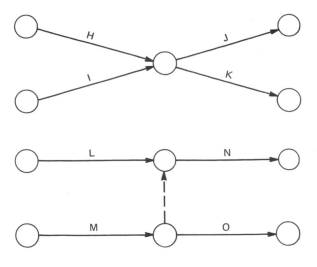

Both activities H and I must be completed before either activity J or K can start.

Both activities L and M must be completed before activity N can start; however, the start of activity O depends upon the completion of activity M alone. The dashed-line arrow is considered a dummy activity, having no duration and used to show the restriction.

Another use of the dummy is eliminating the possibility of two concurrent activities from starting and ending at the same nodes. It would be wrong to show two concurrent activities P and Q thus:

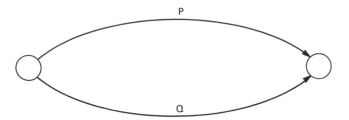

Activities P and Q would be identified by the same nodes. A dummy activity must be introduced to prevent this, thus

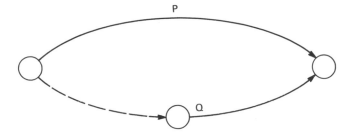

A dummy activity then is used for three purposes: to draft the network in consonance with construction procedure, to facilitate computations by eliminating identical node numberings for two or more

activities, and to incorporate company policy into the network.

The term restriction has been introduced in the ground rules and needs elaboration. We already have seen that a dummy can be used to restrict the start of activity N to the completion of activities L and M. There are three types of restrictions to consider in critical path scheduling. First, there are physical restrictions; for example, the walls of a house cannot be raised until the foundation is completed and the plumbing piping cannot be topped out until both the roughing in and roof sheathing are completed. Second, there are restrictions of company policy and capability; for example, two activities could be performed simultaneously, each requiring the use of a special piece of equipment; but the company owns only one piece of this equipment, so a decision has to be made as to which activity shall be done first. The other activity is restricted to completion of the first one because of equipment limitation. The same restriction could apply because of manpower limitations. The third type of restriction, procurement of materials, subcontracts, and shop drawings, involves a time element and therefore is considered as an activity of its own having a definite duration.

Restrictions can be read into the diagram by sequencing of physical activities, by use of dummies having no duration, or by activity type arrows for time-consuming services.

The complete network should be drafted before the nodes are numbered. As stated previously, a node may be considered as an event, which is the completion of a preceding activity and/or the start of a following activity. It is merely a vehicle used to express the occurrence of the completion or start of an activity. Also, an activity can be identified by two node numbers; the first number is the node at the tail of the arrow, the second number is the node at the head of the arrow.

Activity B now can be identified as activity 1-3 and activity A can be identified as 3-5. The tail of an arrow is at its i node and the head of the arrow is at its j node, so that all arrows are identified by their i-j numbers. The numbers need not be consecutive, but the j number of any activity must always be larger than its i number. In the same vein, no two arrows, including dummies, may have identical i-j numberings. Identical i-j numberings for two activities would confuse the calculations of starting times and finishing times regardless of whether the calculations are manual or computerized.

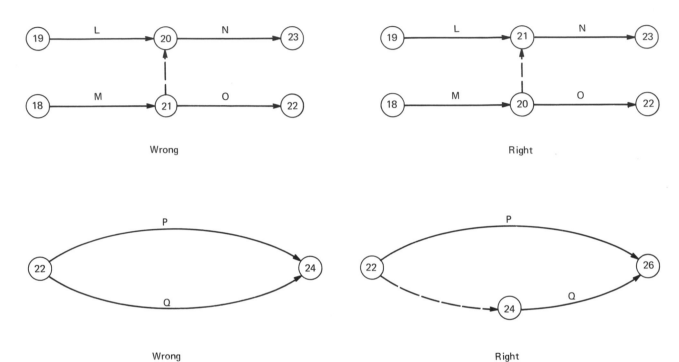

CHAPTER 4

STAGE I
OF
CPM SCHEDULING

CONSTRUCTION OF THE
ARROW DIAGRAM

There are four stages in CPM. Stage I is the construction of the arrow network diagram. Stage II is the allocation of manpower, resources, and time. Stage III is the determination of project duration, the critical activities, and the float times of noncritical activities. Stage IV is the involvement of costs and shortening of project duration. This chapter is concerned with the first stage.

The basis for progress scheduling by any method lies in the preparation of an accurate quantity survey, and the critical path method is no different in this respect. A quantity survey must have been prepared in order that a cost estimate may be determined. The cost estimate, as will be seen later, is based upon the builder's most economical operations for performing the work, or else his competitor would have had a lower bid and received award of the contract. This is normal procedure in contract construction, so this chapter will consider only normal costs and durations. However, a quantity survey, having been prepared primarily for a cost estimate, will have to be modified for construction scheduling. The quantity survey does not show items such as moving in, except as an overhead expense, neither does it show such contract requirements as preparation of shop drawings, and their submission to and approval by the architect, prior to fabrication and delivery of materials. All of these need to be considered in the preparation of a construction progress schedule. Then, too, there are items of cost appearing in the quantity survey that need not be shown in the progress schedule, such as laboratory design of concrete mix, or preparing cylinders of concrete for testing purposes, or rough hardware. The quantity survey would show one lump-sum entry for each activity to be performed by subcontract, whereas the schedule would need to break it down into two or more activities in consonance with construction procedure and individual job requirements.

The estimator who made the quantity survey has a good understanding of the job requirements and it should be his responsibility at this stage to prepare an arrow network diagram for the project. Again using the small warehouse as an example, we are ready now for Stage I of critical path scheduling—the preparation of the arrow network diagram.

The estimator now will make a rough draft of an arrow network diagram based upon his quantity survey. He applies the three principles for the construction of an arrow in the diagram.

The restrictions to consider in sequencing activities are threefold. First, the physical restrictions, such as masonry, cannot be physically installed until the concrete has been placed and cured. Second, job requirements and specifications, for example: even though concrete could physically be placed without installation of reinforcing steel, this job requires its installation before the placing of concrete. The specifications would require that the installation of reinforcing steel be inspected by the architect/engineer before the pouring of concrete, but in order to simplify the demonstration of an arrow network diagram, inspection has been omitted.

Third, company policy places certain restrictions upon activity sequencing. The policy of this company is not to start any interior work until the structure is completely enclosed (i.e., weatherproof).

Dummy activities are indicated by a broken line. They are used to show two differing situations. One disadvantage of the arrow network diagram is that in the numbering of the nodes only one activity can have the same i-j designations. That is to say, no two activities can have the same i-j numbering. Otherwise a computer would reject the system or, in noncomputer approach, the manual calculations for project duration, critical path, and floats would be defeated. Another situation requiring the use of a dummy activity is showing proper sequencing in greater detail. Where the installation of plumbing pipes cannot be installed until after the roughing in, masonry, and roofing have been completed, dummy activities can be introduced from roughing in to pouring and finishing concrete and from roughing in to piping and from roofing to piping.

When the arrow network diagram has been established, the next step is to number the nodes. The only rule to observe is that the j node of any arrow must be numerically greater than the i node. It does not necessarily have to be the consecutively next larger number. In fact, it would be a good idea to skip one or two numbers in order to introduce new activities such as ordering out materials or subcontracts without having to erase and change the numbering of subsequent activities.

To demonstrate this technique, the small warehouse is used once again. The first item of work listed on the quantity survey, Figure 1-2, is job layout. This can then be expressed as activity A,

Now apply the three fundamental questions to this activity.

Question: What activity must immediately precede the start of this activity?

Answer: Move in.

So activity B, move in (or mobilize, as some people call it), is placed immediately preceding activity A.

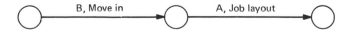

Question: What activity must immediately follow this activity?

Answer: Excavation.

So excavation becomes activity C immediately following activity A.

Note that letter identification of activities need not be in alphabetical order.

Question: What activity may be performed simultaneously with this activity (still referring to activity A)?

Answer: To make up forms, so this becomes activity D.

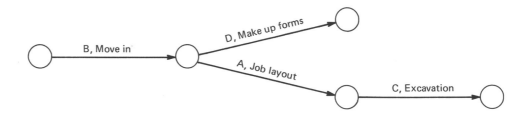

Now with respect to activity B, move in, in answer to the first question one might answer "decision to start" or "notice to proceed, " and it would be perfectly all right to introduce such an activity, but in order to simplify diagramming it is not done at this time. The second question already has been answered in the drafting of activity A. In answer to the third question with regard to activity B, a gamut could be introduced, if not in activities then at least in job restrictions such as ordering of materials for subsequent activities, award and notification of subcontracts or, perhaps of greater consequence, preparation of shop drawings, their submission to and approval by the architect, and their fabrication and delivery to the job site. Very often the time required for shop drawings and the like can be so long as to be critical to the execution of the job. In actual practice, every such restriction should be shown on the arrow diagram, necessitating many activities beginning at node 1. For simplicity, only one such restriction will be shown, as activity E—shop drawings, approval, fabricate and deliver reinforcing steel—and it is drafted to show concurrence with activities B and D.

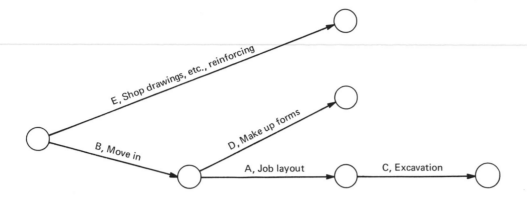

Now apply the three fundamental questions to activity C, excavation.

What must immediately precede it? Activity A, job layout.

What must immediately follow it? Activity F, erection of forms.

What may be performed concurrently? Activity D, make up forms.

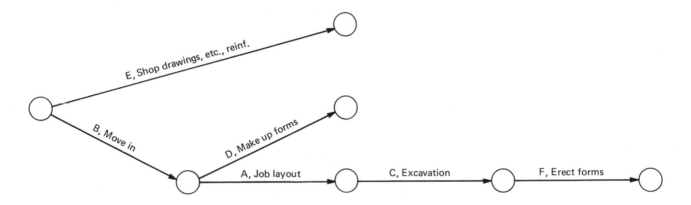

Now apply the three fundamental questions to activity D. It must be preceded by activity B; it must be followed by activity F; and it may be performed concurrently with activities A and C. This could be shown on the diagram as a dummy activity between activities D and F, but it would be unnecessary because the same reading can be achieved by erasing the node at the head of the arrow under activity D and extending the arrow to the tail of activity F, as shown in the full arrow network diagram for this small warehouse, Figure 4-1a.

A word of caution is necessary here. Those who have mastered the logic of CPM tend to exaggerate use of it. There is almost no limit to how many activities can be conjured by one who is inexperienced in building construction. The total number of activities, as well as dummies,

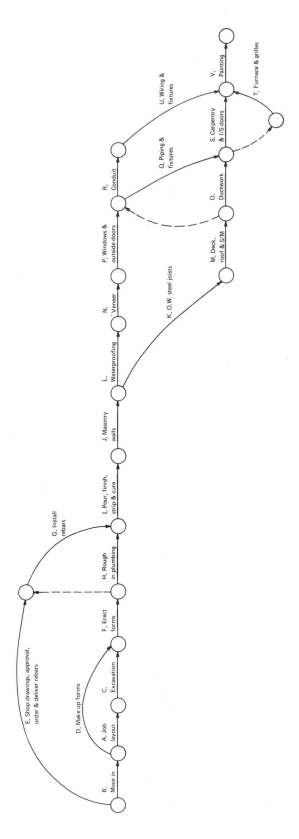

FIGURE 4-1a

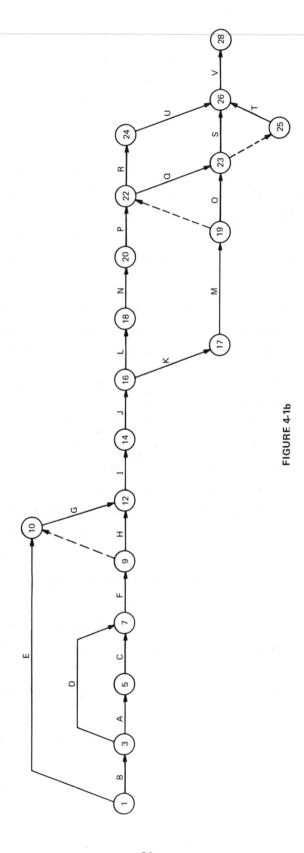

FIGURE 4-1b

33

should be kept to the minimum required to intelligently construct the building. In the erection of formwork, it would be ridiculous to make an activity of driving stakes or, in cement finishing, to make an activity of procuring trowels.

The arrow network diagram with all the nodes numbered, Figure 4-1b, completes Stage I of CPM for the small warehouse.

Before Stage II is begun, advice is needed about the selection of activities and fineness of detail to which they need be considered. In a first rough draft without a quantity survey, only the major divisions of a house could be shown, as in Figure 4-2a. The five major divisions of a house are foundation, framing, sheathing, exterior finish, and interior finish. Scheduling this presents no problems, no complexities. All five divisions are critical. Next, take each division in turn, breaking it down into its components. The foundation could be broken down into batterboards, excavation, forming, reinforcing, inspection, pouring and finish, and stripping and curing. In the case of a slab-on-ground, roughing in of plumbing and plumbing inspection would have to be included, as in Figure 4-2b.

Framing could be broken down into prefabrication of wall sections and trusses and their delivery to job site; erection of walls, joists, and trusses, and piping and wiring.

Sheathing in this instance would be only the roof sheathing because the wall components already would have been sheathed at the mill before delivery to the job. However, plywood for roof sheathing should be pre-cut and each sheet should be numbered in accordance with a shop drawing so that most of the waste would be eliminated.

Exterior finish could be broken down into installation of windows and exterior doors, siding, facia frieze and plancher, shutters, columns, and painting. Interior finish could be resolved into dry wall or plaster, doors and trim, cabineting, flooring, and the rest.

The tendency is to overdetail. In Figure 4-2b, it is not really necessary to make a separate activity of "make up crow's foot." This could be considered as an integral part of "rough-in plumbing." There is no limit to the details into which each activity could be divided. It would be ridiculous, for example, to break down the wall framing into plates, studs, door bucks, window bucks, corners, tees, wind bracing, and fire-stops.

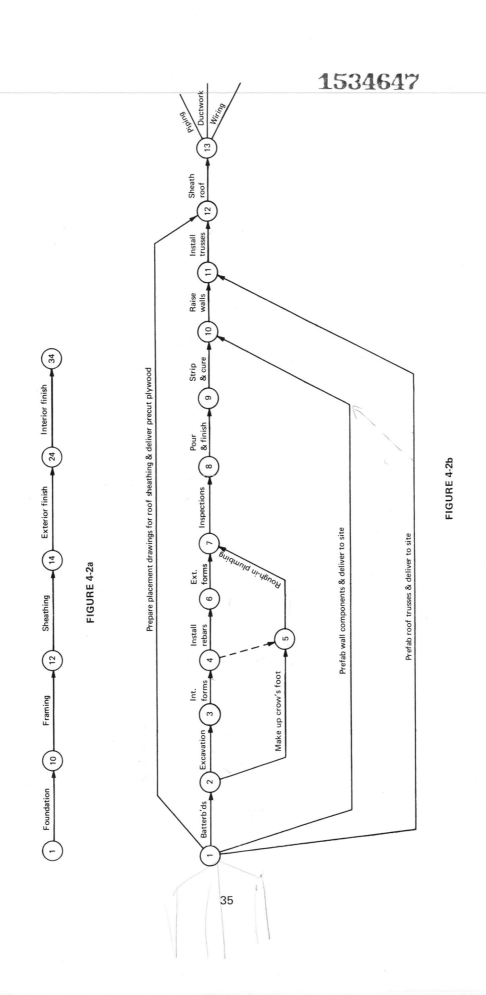

FIGURE 4-2a

FIGURE 4-2b

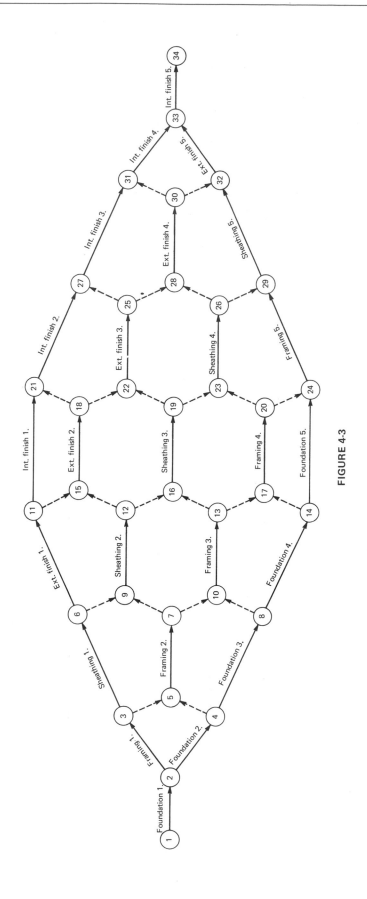

FIGURE 4-3

Subcontract work need be scheduled only from where the general contractor has to be ready for it to where it needs to be completed so that a following activity may begin. The subcontractor should plan and prepare his own subchain of activities very well in order to fit in with the master project schedule.

Where a group of houses is planned, the same network logic would apply if the houses were assigned numbers and each activity would carry the house number with it. In the over-all planning and scheduling of a group of houses, probably only the five major parts of the house, as scheduled in Figure 4-2a, need be considered.

Figure 4-3 shows the drafting of an arrow network for five houses, using the five major parts. The same procedure could be used for any type of project whose work is of a repetitive nature. With little trouble, Figure 4-3 could be elaborated to show all of the work components of all five houses. Ascribing time durations to each activity and determining the critical path and float times could be of great assistance in planning the most economical use of resources.

This completes Stage I of CPM.

CHAPTER 5

STAGE II
OF
CPM SCHEDULING

ALLOCATION OF MANPOWER,
TIME, AND RESOURCES

Stage II is the allocation of manpower, time, and resources. It would be an oversimplification to say that the estimator merely turns over the network diagram to the project superintendent and asks him to assign the time durations to the several activities. This would be the ideal situation if it could be made to work. Very often the superintendent-to-be feels that he is too busy; yet he is the man who has to study the plans, obtain a general picture of the project, and execute the work within contract time obligations. In a sense, it can be said that the project now has been planned but yet needs to be scheduled. Before the project can be scheduled, each activity must be analyzed with respect to the quantity and nature of the work to be performed and then integrated with the available manpower, equipment resources, and material deliveries in order to determine the duration of each activity.

Of course, the best determination of activity duration can be made by the superintendent, who bases it on his previous experience, but he is not always readily available for this. It has been said that up through this stage the greatest effect of critical path scheduling has been to force management to analyze the job prior to construction. This is true. However, because most building construction projects are bounded by a time limit, it is necessary that the time boundaries be determined now, prior to start of construction. So the estimator now must make these determinations. But he is not entirely unaided. He has before him the quantity survey and the pricing-out sheets from which he can make a reasonable determination of activity durations. Basically this procedure in backtracking consists of dividing the total cost of labor of each activity by the cost per day per crew. The quotient is the number of crew days estimated to be required to perform the activity. Should the activity require the use of equipment, then the rental rate of the equipment would be included in the crew cost per day.

The cost estimate, Figure 1-2, and the network diagram, Figure 4-1, will be used to demonstrate how an estimator could determine activity durations.

Activity A, job layout, shows a total cost of $90, of which $45 is labor. A carpenter and a laborer earn $56 per day. This means then that they could set up batterboards and lay out the job in a little less than one eight-hour day. So the duration to be ascribed to activity A would be one day.

Activity B, move in, can be done easily in one day on this small a job.

Activity C, excavation, carries a total cost of $500, which does not include materials. Activity C would be performed by use of a trenching machine plus two laborers dressing-out at a cost per day of $182, requiring two and three-quarter days to perform, so the duration would be three days.

Activity D, make up forms, contains a labor cost of $1,200. This work would be performed by a crew of four carpenters and two laborers at a crew cost of $200 per day, giving a duration of six days.

Activity E, shop drawings, approval, order, and delivery of reinforcing, can be determined to require ten days simply by a query to the steel company.

Activity F, erection of forms, has a labor cost of $1,220 at the same crew cost per day as make-up of forms. It requires a duration of six days.

Activity G, install rebars, would have a labor cost of $240 and a crew cost (three rodbusters) of $120 per day or two days' duration.

Activity H, rough-in plumbing, by telephone call to the mechanical subcontractor, is determined to require three days.

Activity I, pour, finish, strip, and cure concrete, can be estimated to have a labor cost of $1,367, but the duration would not be determined by crew cost because the concrete must be poured and finished in one day. Its duration would be determined by the specified time for curing, usually one calendar week or five days.

Activity J, masonry walls, does not include the brick veneer because the block walls have to be erected first and then waterproofed before the veneer can be installed. The labor cost for this activity is $3,800— a crew could consist of one foreman, five masons, and six laborers at a crew cost of $380 per day, requiring a duration of ten days.

Activity K, open web steel joists, would require a minimum of one piece of hoisting equipment with operator, two laborers, and two steel erectors at a crew cost of $240 per day, which when divided into an erection cost of $1,200 gives a five-day duration.

Activity L, waterproofing, would have a labor cost of $600 and a crew cost of $120 per day, or five days' duration.

Activity M, roof decking, roofing, and sheet metal, would require four days as determined by a call to the roofing subcontractor.

Activity N, masonry veneer, using the same crew that was used for activity J, masonry walls, requires a five-day duration.

Activity O, ductwork, is an eight-day job, as determined by the mechanical subcontractor.

Activity P, windows and exterior doors, is estimated to require two days of work by two carpenters.

Activity Q, piping and mechanical fixtures, would require the subcontractor three days.

Activity R, installation of electrical conduit, would require three days.

Activity S, carpentry and interior doors, can be determined to require three days. The rough carpentry consists of 800 BF and one carpenter reasonably can be expected to fit and install about 450 BF of lumber and eight interior doors per day.

Activity T, furnace and grilles, and activity U, wiring and fixtures, have durations determined by the mechanical and electrical contractors of two and four days, respectively.

Activity V, painting, is the last activity to be scheduled. It is proverbial in construction circles that "the painter is the last man off the job." He has to cover up the mistakes made by all other tradesmen. The estimate for painting labor is about $1,200. The daily cost per crew of three painters is $120, giving a ten-day duration for painting.

The next step in Stage II is to place each activity duration below the activity description of each arrow in the network—see Figure 5-1.

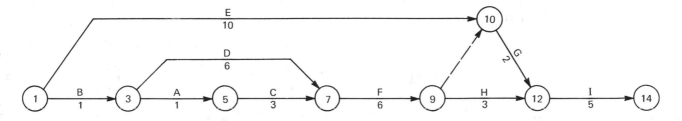

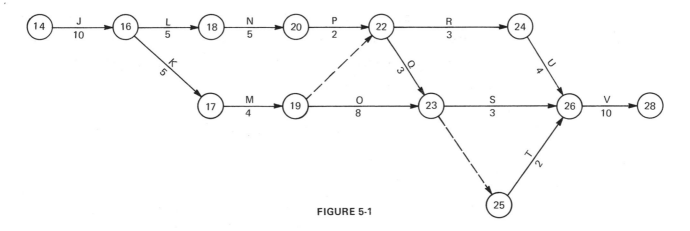

FIGURE 5-1

This completes the allocation of time and resources for the small warehouse project.

In house carpentry, individual items of work such as studs, rafters, and so on, vary slightly from an average rate of production of 450 BF per eight-hour day.

On large projects, a difficulty in planning and scheduling arises when certain activities interrelate to the extent that a following activity can commence before a preceding activity is entirely completed. An example is the case in which the foundation for a large building could be built in two or more sections. In such a case, the formwork on the first section could be performed simultaneously with the excavation of the second section, and the installation of reinforcing steel and the pouring of concrete could be performed simultaneously with the formwork on the second section, as shown in Figure 5-2. This technique is all right when various sections of an activity can be specifically identified and time and resources allocated properly. But when activities do not lend themselves to recognizable sections, it is poor practice to attempt to assign arbitrary sections such as activity W, start steel erection, activity X, continue steel erection, and activity Y, finish steel erection. This might have been done in order to schedule an activity of start masonry after the steel erection has been started and concurrent with the continuation of steel erection, as shown in Figure 5-3. Because the erection of both the steel and masonry walls are continuous items of work, it would be more in consonance with their nature and would better reflect job conditions to introduce a time-lapse dummy activity of lead time. In other words, after sufficient steel has been erected, the masonry crew could begin laying masonry following up the steel erection. The two activities then are performed concurrently, but the masonry cannot begin until a sufficient amount of steel has been erected. This situation is much better scheduled as shown in Figure 5-4.

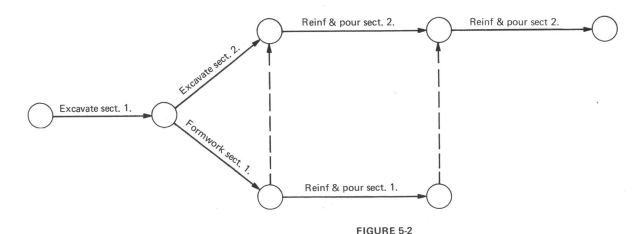

FIGURE 5-2

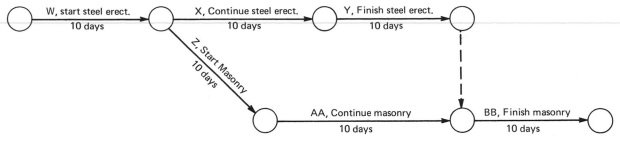

FIGURE 5-3

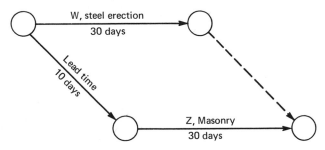

FIGURE 5-4

CHAPTER 6

STAGE III
OF
CPM SCHEDULING

CALCULATIONS
ALONG THE ROUTE

Stage III of CPM consists of the determination of the project duration, the critical path, and the float time of the remaining noncritical activities.

The first step in Stage III is to fill out form CPM 1 through the columns headed Duration—see Figure 6-1. Note that to indicate activity codes, letters of the alphabet are used, rather than the decimal system used on the cost estimate sheet. The letters are not preferred by the author because most projects require so many activities that they go into doubled and tripled lettering. A million-dollar building project usually requires some 300 activities. Then, too, it is just as convenient to refer to activities, in the arrow network diagram, by their i-j identification; for example, erection of forms would be identified as activity 7-9. A dummy activity is no activity at all, merely a relationship of sequencing. This relationship may affect the project duration and so it must be listed by its i-j identification.

Job: Small Warehouse   Calculations for Float   CPM 1

| Node No. | | Act. Code | Description | Duration | Earliest | | Latest | | Float | |
i	j				Start	Finish	Start	Finish	Total	Free
1	3	B	Move in	1	0	1	0			
1	10	E	Shop drawings - rebars	10		10				
3	5	A	Job layout	1						
3	7	D	Make up forms	6						
5	7	C	Excavation	3						
7	9	F	Erect forms	6						
9	10		Dummy	0						
9	12	H	Rough-in plumbing	3						
10	12	G	Install rebars	2						
12	14	I	Pour concrete, finish & cure	5						
14	16	J	Masonry walls	10						
16	17	K	O.W. steel joists	5						
16	18	L	Waterproofing	5						
17	19	M	Deck, roof, & sheet metal	4						
18	20	N	Brick veneer	5						
19	22		Dummy	0						
19	23	O	Ductwork	8						
20	22	P	Windows & ext. doors	2						
22	23	Q	Piping & fixtures	3						
22	24	R	Elect. conduit	3						
23	25		Dummy	0						
23	26	S	Carpentry & int. doors	3						
24	26	U	Wiring & fixtures	4						
25	26	T	Furnace & grilles	2						
26	28	V	Painting	10						

FIGURE 6-1

In the listing of activities from the arrow network to form **CPM 1,** every arrow in the network, including dummies, must be listed. They do not have to be listed in any particular order, but for facility of calculating starts and finishes, they should be listed in increasing order of the i node of their identification.

The estimator should assume that the durations indicated by the superintendent are faithful; that given the men, equipment, and timely delivery of materials, these durations are the most economical for his performance. However, it is proper to question some durations, because many superintendents erroneously believe that by showing a greater duration for a specific activity they will gain in the opinion of the contractor when he completes this activity in less time than scheduled. When the duration of an activity has been determined at its most economical cost, a lessening of its duration can be achieved only through the expenditure of additional money on overtime, or double shifting, or additional equipment at less efficiency. Occasionally a superintendent has lost his position with a contractor for completing a project ahead of schedule—he dissipated the profit by speeding up operations.

Satisfied that all the durations assigned by the superintendent are in order, the estimator now takes the arrow network and form CPM 1 and begins his calculations to determine the project duration by computing the early start times and early finish times of each activity. This feature of the work may be performed in one of three different ways, viz., by computer, which requires the input of only the i-j designations and the duration of each activity; by the determination of the earliest possible occurrence, EPO, of each node in the network, as shown by a number above each node in Figure 6-2; or by completing the dual column of earliest start and finish on CPM 1, as shown in Figure 6-3.

In either of the three methods, all starts and finishes are considered as of the end of that day. For example, activity 1-3, which starts the job concurrently with activity 1-10, would start at the end of zero day and finish at the end of one day. So, too, activity 1-10 would start at the end of zero days and finish at the end of ten days.

Three techniques for determining project duration and floats have been mentioned. The computer method will be mentioned in a later chapter. The method of nodes is demonstrated in Figure 6-2. The method

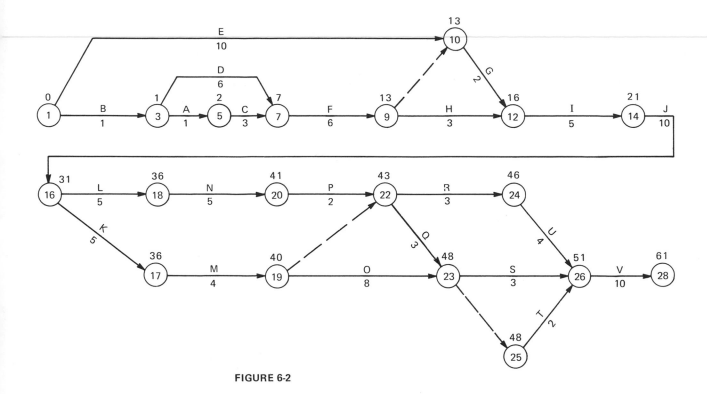

FIGURE 6-2

of arrows is demonstrated in Figure 6-3. Both will be discussed in detail now.

The method of nodes uses only the arrow network diagram. Because a node is the indication of an event, it is an occurrence in the diagram. A node indicates the completion of a preceding activity and the start of a following activity. This method then works with the earliest possible occurrence (EPO) of an event such as the start of an activity and the latest possible occurrence (LPO) such as the completion of an activity. EPOs are entered above the node pointing forward to the completion of the project. LPOs are entered below the node, reading backward toward the start of the project.

In Figure 6-2, the number zero is placed above node 1. The EPO of all activities starting at node 1 is at the end of the zero day. Node 3 occurs at the completion of activity B, move in, whose duration is one day. Adding this one-day duration to the EPO of node 1 gives the finish time of activity B as of the end of the first day, which becomes the EPO of node 3. Two activities have their beginning at this node, activity A, job layout and activity D, make up forms. Activity A has a duration of

Job: Small Warehouse   Calculations for Float   CPM 1

| Node No. | | Act. Code | Description | Duration | Earliest | | Latest | | Float | |
i	j				Start	Finish	Start	Finish	Total	Free
1	3	B	Move in	1	0	1				
1	10	E	Shop drawings - rebars	10	0	10				
3	5	A	Job layout	1	1	2				
3	7	D	Make up forms	6	1	7				
5	7	C	Excavation	3	2	5				
7	9	F	Erect forms	6	7	13				
9	10		Dummy	0	13	13				
9	12	H	Rough-in plumbing	3	13	16				
10	12	G	Install rebars	2	13	15				
12	14	I	Pour concrete, finish & cure	5	16	21				
14	16	J	Masonry walls	10	21	31				
16	17	K	O.W. steel joists	5	31	36				
16	18	L	Waterproofing	5	31	36				
17	19	M	Deck, roof, & sheet metal	4	36	40				
18	20	N	Brick veneer	5	36	41				
19	22		Dummy	0	40	40				
19	23	O	Ductwork	8	40	48				
20	22	P	Windows & ext. doors	2	41	43				
22	23	Q	Piping & fixtures	3	43	46				
22	24	R	Elect. conduit	3	43	46				
23	25		Dummy	0	48	48				
23	26	S	Carpentry & int. doors	3	48	51				
24	26	U	Wiring & fixtures	4	46	50				
25	26	T	Furnace & grilles	2	48	50				
26	28	V	Painting	10	51	61				

FIGURE 6-3

one day, and it ends at node 5. So the EPO of node 5, being the sum of the EPO of node 3 and the duration of activity A, is two days. Two days then is the earliest possible starting time for activity C, excavation, just as one day is the earliest possible starting time for activity D. Both activities end at node 7. The EPO of node 7 would be the later or greater of the two earliest possible finish dates. The EPO of activity C at two days plus its duration of three days adds up to a possible occurrence of five days for node 7. The alternative of the EPO of node 3 at one day plus the six-day duration of activity D gives a possible occurrence of seven days at node 7. So the EPO of node 7 is at the end of the seventh working day. The earliest possible starting date for activity F, erect forms, is the end of the seventh day, its duration is six days, and it ends at node 9, so the EPO of node 9 would be 13. And so on through the network until the EPO of node 28 is reached at the end of the 61st working day. This small warehouse has a project duration, at normal costs, of 61 working days.

The method of arrows arrives at the same project duration and perhaps may be preferable because it is a simple, routine process of arithmetic, not requiring use of the diagram except for reference purposes. The arithmetic work involved in the method of arrows is a continuation process on form CPM 1. Form CPM 1 has been prepared previously and lists every arrow, including those for dummy activities through the column headed "Duration." The estimator now completes the dual column headed "Earliest" to determine the project duration (see Figure 6-3).

He starts out by entering a zero in the start column on every row where the figure 1 appears in the i row. Adding the duration of each activity to the figure in the start column gives him the figure to enter in the finish column. In Figure 6-3, it can be seen that 1 appears in the i column twice, just as there are two arrows leaving node 1 on the diagram. So 0 is entered in the start column in each of the two rows where 1 appears in the i column. Adding the duration of one day for activity B to the 0 gives 1, to be entered in the finish column. On that same row of activity B, 3 appears in the j column, so wherever 3 appears in the i column the figure 1 is entered in the start column of the same row, as was done in the rows of activities A and D. It now can be stated that the start for any activity is the finish of some previous activity where the i number of the starting activity is the same as the j number of the finishing activity.

In those instances where there is more than one finishing activity with identical j numbers but differing finish times, such as for activities C and D, the greater finish time of the several activities must be used as the start time of the starting activity. Because C finished at the end of day 5, and D at the end of day 7, the start time of activity F would be at the end of day 7. By continuing the process through to completion, it is found that the last activity in the project, activity V, has an early finish time of 61 days and once again the project duration has been determined to be 61 working days.

It appears to be wise at this stage to compare the project duration at the most economical costs to the contract time for completion of the project, converting working days to calendar days, if necessary, and allowing for any lost time due to rain or holidays. If the project duration falls within the contract time, well and good. But if the project duration is greater than the contract time, then the contract becomes subject to the time delay damages clause. If the project duration can be shortened to fall within the contract time of completion, it can do so only by shortening the duration of one or more activities. To shorten the duration of an activity increases its cost, because it requires either overtime, double shifting, more expensive materials and/or methods, more men (thereby decreasing the efficiency of the crews), or additional equipment (requiring the outlay of operating capital). Then too the additional expense of shortening one activity will vary from the additional expense of shortening others. Also to be considered is the fact that shortening an activity may not necessarily shorten the project duration. For example, in Figure 6-2, shortening the duration of activity K—16-17, OW steel joists—from five to two days, if at all possible, would not shorten the duration of the project because activities 16-17 and 17-19 required nine days of work during the same occurrences of nodes 16 and 22. It now is apparent that some activities control the length of project duration whereas others do not. It behooves the superintendent of construction to distinguish between the two types.

In every construction project there is no time at which a project need stand still. Another way of making this statement is: In any and every project something always can be done. With the exception of the final activity, some activity always can begin upon the completion of another activity, without any delay of time. There must be at least one continuing chain of activities from beginning to end of a project. There may be more than one chain, or there may be two or more subchains, but there must be at least one. Those activities forming one or more chains or sub-chains continuously are the ones called critical activities. Those activities that do not control the project duration are noncritical and, therefore, the term float may be ascribed to all noncritical activities. Float will be defined and discussed in detail later in this chapter.

The determination of all critical activities is made then by an analysis of float or leeway. Those activities that have no float time may be critical. In other words, critical activities have no float. There are situations in which an individual activity may have no float yet still may not be critical because it is not a link in the chain that must be unbroken from start to finish of the project.

Again, there are two methods for the determination of float times or leeway, the method of nodes and the method of arrows. The method of nodes will be taken up first.

In Figure 6-4, numbers have been placed immediately below the nodes to indicate the LPO (latest possible occurrence) of the node.

The EPO of the last node in the diagram, which determines the project duration, is also the LPO of the last node. To determine the LPO of any node, a subtraction is necessary. Each arrow has two nodes, an i and a j node. The LPO of any i node is the remainder, determined by subtracting the duration of an activity from the LPO of its j node. The LPO of node 26 is 61 minus 10, or 51. The LPO of node 25 is 51 minus 2, the duration of activity T, furnaces and grilles, or 49. The LPO of node 24 is 51 minus 4, the duration of activity U, wiring and fixtures, or 47.

Wherever a node has more than one arrow leading away from it, such as node 23, there is more than one possibility for its LPO. Each must be considered and the least possibility must be used. Thus, backing up from node 26, and an LPO of 51, a duration of three days, activity S, gives an apparent LPO of node 23 of 48 days. Backing up from node 25, and an LPO of 49 days, a duration of zero, a dummy activity, gives an apparent LPO of node 23 of 49 days. The lesser of the two possibilities,

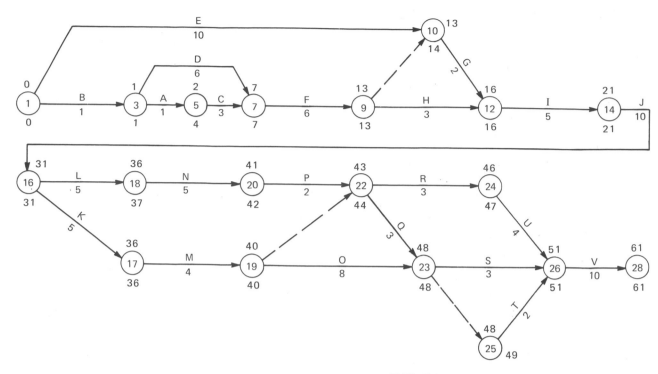

FIGURE 6-4

48 days, must be the LPO of node 23, and is so entered on the diagram, immediately below it. And so on until node 1, the start of the project, is reached with the figure 0. If a number other than zero is reached at node 1, then an arithmetic error has been made; the computations must come back to time zero at the start of the project.

Upon examination of the various nodes, it will be seen that some have identical EPOs and LPOs and some have LPOs numerically greater than their EPOs. Under no circumstance can a node have a greater EPO than LPO. The difference between the LPO and EPO is called the leeway of the node.

Leeway is defined as the amount of time that could exist between the latest and earliest possible occurrence of a node. At node 5, where the EPO is 2 and the LPO is 4, there are two days of leeway between the completion of activity A, job layout, and the start of activity C, excavation.

In consonance with a foregoing statement that there always is something that can be performed on a construction project, it is only logical that there should be at least one chain of events that have no leeway. In Figure 6-4 it can be seen that this chain lies along the path of nodes 1,

3, 7, 9, 12, 14, 16, 17, 19, 23, 26, and 28. This is the critical path, and is so indicated by placing double slash marks immediately after the node, across the arrow leading to the next node with no leeway (see Figure 6-5).

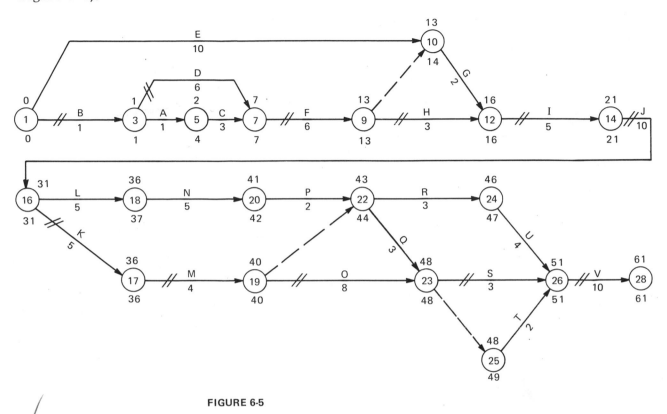

FIGURE 6-5

The method of arrows makes the determination of the same critical path by means of a series of arithmetic subtractions using the dual column of latest starts and finishes and the column of total float, all on form CPM 1, as demonstrated in Figure 6-6. Starting on the row of activity V, painting—the last activity in this project—the project duration of 61 days is entered in the column of latest finishes. Subtracting the duration of activity V, 10 days, from its latest finish, 61 days, gives its latest start, 51 days. Because the i node of activity V is 26, all activities immediately preceding it must have a j node of 26. In the j column, the figure 26 appears three times, lying in the rows of activities T, U, and S. In each of these three rows, the figure 51 is entered under the column of latest finish. The duration of each of the three activities is subtracted from its latest finish to give its latest start.

Job: Small Warehouse   Calculations for Float   CPM 1

| Node No. | | Act. Code | Description | Duration | Earliest | | Latest | | Float | |
i	j				Start	Finish	Start	Finish	Total	Free
1	3	B	Move in	1	0	1	0	1		
1	10	E	Shop drawings - rebars	10	0	10	4	14		
3	5	A	Job layout	1	1	2	3	4		
3	7	D	Make up forms	6	1	7	1	7		
5	7	C	Excavation	3	2	5	4	7		
7	9	F	Erect forms	6	7	13	7	13		
9	10		Dummy	0	13	13	14	14		
9	12	H	Rough-in plumbing	3	13	16	13	16		
10	12	G	Install rebars	2	13	15	14	16		
12	14	I	Pour concrete, finish & cure	5	16	21	16	21		
14	16	J	Masonry walls	10	21	31	21	31		
16	17	K	O.W. steel joists	5	31	36	31	36		
16	18	L	Waterproofing	5	31	36	32	37		
17	19	M	Deck, roof, & sheet metal	4	36	40	36	40		
18	20	N	Brick veneer	5	36	41	37	42		
19	22		Dummy	0	40	40	44	44		
19	23	O	Ductwork	8	40	48	40	48		
20	22	P	Windows & ext. doors	2	41	43	42	44		
22	23	Q	Piping & fixtures	3	43	46	45	48		
22	24	R	Elect. conduit	3	43	46	44	47		
23	25		Dummy	0	48	48	49	49		
23	26	S	Carpentry & int. doors	3	48	51	48	51		
24	26	U	Wiring & fixtures	4	46	50	47	51		
25	26	T	Furnace & grilles	2	48	50	49	51		
26	28	V	Painting	10	51	61	51	61		

FIGURE 6-6

It should be noted that this process gives two separate latest starts for activities beginning at node 23. The latest start for activity S being 48 days and the latest start for the dummy activity being 49 days, the least or earliest of the two latest starts must be used for the latest finish of all activities ending at that node. In the example, two activities, O and Q, end at node 23 so 48 days is the latest finish time for both of these activities and 48 is entered under the column of latest finish. The two rows can be found by looking in the j column for node 23. This procedure is continued until completed, coming back to node 1 with a latest start time of zero. Once again the process must come back to time zero; if not, an error in arithmetic has been made.

The next step is the determination of total float.

Total float is defined as the amount of time the duration of an activity can be lengthened or the amount of time its start can be delayed, without affecting the duration of the project.

On form CPM 1, Figure 6-7, total float can be determined in any one of three ways, expressed by the following formulas:

(1) $TF = LF - EF$

(2) $TF = LS - ES$

(3) $TF = LF - ES - Duration$

where: TF is total float, LF is latest finish time, EF is earliest finish time, LS is latest start time, ES is earliest start time.

The value to the superintendent of knowing total float times is obvious. Where two or more activities fall in the same subchain, each may have an identical amount of total float. This does not mean that the total float in each activity is additive. The total float is selective to any one of the activities in the subchain, but no more than one. Once used for one activity, it is lost to all of the others in that particular subchain.

All activities on the critical path must have zero total float. The converse is not necessarily true. There may be one or more activities with zero total float that are not on the critical path. The critical path, however, is determined by at least one continuous chain of activities with zero total float from start to finish of the project. Figure 6-7 shows such a chain. By looking down the column of total floats and marking those activities that have zero total floats, one can see that one chain of

Calculations for Float

Job: Small Warehouse

CPM 1

| Node No. | | Act. Code | Description | Duration | Earliest | | Latest | | Float | |
i	j				Start	Finish	Start	Finish	Total	Free
1	3	B	Move in	1	0	1	0	1	0	
1	10	E	Shop drawings - rebars	10	0	10	4	14	4	
3	5	A	Job layout	1	1	2	3	4	2	
3	7	D	Make up forms	6	1	7	1	7	0	
5	7	C	Excavation	3	2	5	4	7	2	
7	9	F	Erect forms	6	7	13	7	13	0	
9	10		Dummy	0	13	13	14	14	1	
9	12	H	Rough-in plumbing	3	13	16	13	16	0	
10	12	G	Install rebars	2	13	15	14	16	1	
12	14	I	Pour concrete, finish & cure	5	16	21	16	21	0	
14	16	J	Masonry walls	10	21	31	21	31	0	
16	17	K	O.W. steel joists	5	31	36	31	36	0	
16	18	L	Waterproofing	5	31	36	32	37	1	
17	19	M	Deck, roof, & sheet metal	4	36	40	36	40	0	
18	20	N	Brick veneer	5	36	41	37	42	1	
19	22		Dummy	0	40	40	44	44	4	
19	23	O	Ductwork	8	40	48	40	48	0	
20	22	P	Windows & ext. doors	2	41	43	42	44	1	
22	23	Q	Piping & fixtures	3	43	46	45	48	2	
22	24	R	Elect. conduit	3	43	46	44	47	1	
23	25		Dummy	0	48	48	49	49	1	
23	26	S	Carpentry & int. doors	3	48	51	48	51	0	
24	26	U	Wiring & fixtures	4	46	50	47	51	1	
25	26	T	Furnace & grilles	2	48	50	49	51	1	
26	28	V	Painting	10	51	61	51	61	0	

FIGURE 6-7

activities forms the critical path. That chain consists of activities B, 1-3; D, 3-7; F, 7-9; H, 9-12; I, 12-14; J, 14-16; K, 16-17; M, 17-19; O, 19-23; S, 23-26; V, 26-28. This chain of critical activities—or as it is now known, the critical path—can be indicated on the arrow network diagram in four ways.

(1) By shading the arrow heavier than activities containing float.

(2) By drawing double slash marks across the arrow.

(3) By superimposing a wavy line over the arrow.

(4) By double-lining the arrow.

The last method is preferable and is used in Figure 6-8 to indicate the critical path for the small warehouse. The double arrow is continuous from beginning to end.

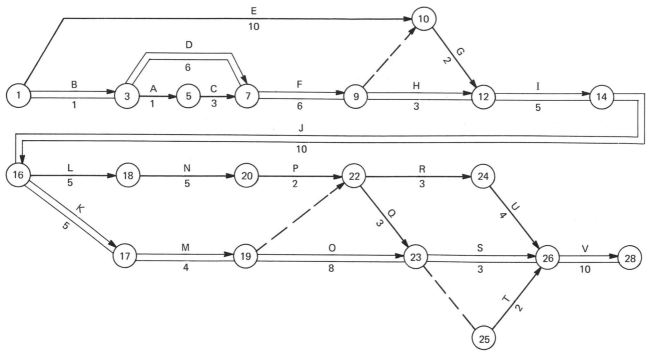

FIGURE 6-8

Total float already has been formulated and defined as the amount of time the duration of an activity can be lengthened or its start delayed without affecting the project duration. Free float now can be defined as the amount of time the duration of an activity can be lengthened or its start delayed without affecting the early start of any other activity. The free float of any activity can be determined by use of the following formula:

$$FF(1) = ES(2) - ES(1) - Dur~(1)$$

where: FF(1) is the free float of any activity

ES(2) is the earliest start of the following activity

ES(1) is the earliest start of the activity itself,
                and

Dur(1) is its duration

In Figure 6-9, the first activity having total float is activity 1-10, with 4 days of total float, ES is 0 and duration is 10. The following activity is 10-12, ES is 13; 13 minus 10 minus 0 gives 3 days of free float to activity 1-10.

The next activity having total float is 3-5, with two days of total float, ES is one and its duration is one day. The following activity is 5-7, ES is two; 2 minus 1 minus 1 is 0 days of free float to activity 3-5.

The next activity having total float is 5-7, with two days, ES is 2 and duration is three days. The following activity is 7-9, its ES is seven

Job: Small Warehouse

Calculations for Float

CPM 1

| Node No. | | Act. Code | Description | Duration | Earliest | | Latest | | Float | |
i	j				Start	Finish	Start	Finish	Total	Free
1	3	B	Move in	1	0	1	0	1	0	
1	10	E	Shop drawings - rebars	10	0	10	4	14	4	3
3	5	A	Job layout	1	1	2	3	4	2	0
3	7	D	Make up forms	6	1	7	1	7	0	
5	7	C	Excavation	3	2	5	4	7	2	2
7	9	F	Erect forms	6	7	13	7	13	0	
9	10		Dummy	0	13	13	14	14	1	0
9	12	H	Rough-in plumbing	3	13	16	13	16	0	
10	12	G	Install rebars	2	13	15	14	16	1	1
12	14	I	Pour concrete, finish & cure	5	16	21	16	21	0	
14	16	J	Masonry walls	10	21	31	21	31	0	
16	17	K	O.W. steel joists	5	31	36	31	36	0	
16	18	L	Waterproofing	5	31	36	32	37	1	0
17	19	M	Deck, roof, & sheet metal	4	36	40	36	40	0	
18	20	N	Brick veneer	5	36	41	37	42	1	0
19	22		Dummy	0	40	40	44	44	4	3
19	23	O	Ductwork	8	40	48	40	48	0	
20	22	P	Windows & ext. doors	2	41	43	42	44	1	0
22	23	Q	Piping & fixtures	3	43	46	45	48	2	2
22	24	R	Elect. conduit	3	43	46	44	47	1	0
23	25		Dummy	0	48	48	49	49	1	0
23	26	S	Carpentry & int. doors	3	48	51	48	51	0	
24	26	U	Wiring & fixtures	4	46	50	47	51	1	1
25	26	T	Furnace & grilles	2	48	50	49	51	1	1
26	28	V	Painting	10	51	61	51	61	0	

FIGURE 6-9

days; 7 minus 2 minus 3 gives 2 days of free float to activity 5-7.

In continuing the process of analyzing each activity having total float it can readily be seen that in any subchain where each activity has an equal total float that free float pertains only to the last activity in the subchain. Dummy activity 9-10 has one day of total float, zero days of free float, activity 10-12 has one day of total float and one day of free float. But because activity 10-12 lies also in the subchain of 1-10 and 10-12, it now is understandable why activity 1-10 with four days of total float has only three days of free float.

By the same token, dummy activity 19-22, with four days of total float, would compute to only three days of free float. These three days of free float are peremptorily struck out because a dummy activity cannot have free float. A dummy can have total float but not free float. When a dummy appears to have free float ascribed to it and there are one or more activities with zero free float preceding it in a subchain, the free float of the dummy would be ascribed to the activity in the subchain immediately preceding it.

Ordinarily, this completes the mechanics of Stage III of CPM.

To summarize: The sequence of operations has been planned, durations have been ascribed to the various activities, a critical path has been projected, and float times have been determined. If nothing more has been accomplished, the estimator and superintendent have been forced to examine the project in its entirety.

It is very likely that the superintendent now would replan one or more activities in view of their being critical, or in view of their float. By replanning the use of men and equipment, he now may see where he can shorten an activity without increasing its cost or he may take advantage of float time by lengthening an activity and reducing its cost at the same time. He now may see that he might need a breakdown in greater detail of some activities, or that others have been scheduled in too great detail.

With planning and scheduling of the project now completed, a more effective tool for construction progress control is available.

CHAPTER 7

CALENDAR-TIME PHASING AND UPDATING

Up to now, the graphics of the arrow network diagram have been drawn without regard to time; the length of arrow was drawn to suit only convenience and logic. Its use then, though of significant importance, has been limited to administrative planning. Most building construction contracts specify that the building be completed and ready for owner occupancy by a specified calendar date or in so many calendar days. Furthermore, the contract specifications require that the builder submit for the architect's approval a progress schedule showing how the builder proposes to proceed with the works in order to complete the project in the specified time as well as showing the anticipated starting and completion dates of all significant items of work. Compliance with this clause of a building construction contract is usually secured with a bar chart. It just as surely could be secured with an arrow network diagram if the diagram were calendar-time phased.

Architects, engineers, contractors, subcontractors, foremen, material suppliers, and the rest of the people interested in building construction are familiar with bar charts. All know the failings of a bar chart and conduct their activities accordingly. But a schedule prepared using CPM has a significant impact on their attitudes. It automatically tells them that the planning and scheduling of the project have been performed in greater detail; failure to deliver by a specified date will delay completion of the project and therefore a desire to participate in the teamwork is inculcated.

It is too much to ask them to forget how to read a bar chart and to learn to read an arrow network diagram. However, a calendar-time phased diagram would not be too difficult to understand. Such a diagram can be seen in Figure 7-1.

The first step in setting up a calendar-time phased schedule is to lay out the calendar to scale. The length of the sheet required will depend on the number of calendar days required for the work, taking into account all legal holidays, Saturdays, Sundays, and anticipated days lost due to rain. The width of the sheet will depend on the number of activities that can be performed concurrently and the space wanted to prevent crowding.

The first or top row is reserved for the months, the second row for the days of the month, the third row for consecutive calendar days, and the fourth row for working days. After the calendar and calendar days have been entered but before the working days are entered, all columns headed by Saturdays, Sundays, holidays, and anticipated days lost to rain should be blacked out.

At or near the center of the sheet, all of the critical activities should be laid out in as straight a line as possible. These are the activities that determine the project duration and therefore control the layout of the progress schedule. The only reason why a break in the line could occur, up or down, is if a dummy activity were to be critical. In the laying out of critical activities, the length of arrows would correspond to their number of working days. In the laying out of noncritical activities, only the horizontal projection of the arrows would correspond to their working days. These arrows would have to be drawn on a slant with earliest start and earliest finish times as a straight, slanted line. Inasmuch as the noncritical activity has some float, the arrow will not reach the end node, falling short in working days by the amount of free float. Where two or more noncritical activities in a subchain have total

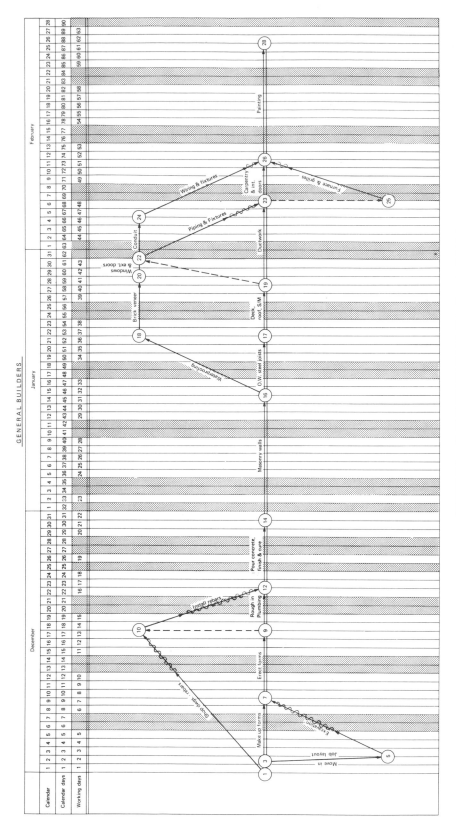

FIGURE 7-1   Small Warehouse Construction Progress Schedule.

float, free float is ascribed to the last activity in that subchain. Refer to activities 3-5 and 5-7. Both have two days of total float, but only 5-7 has two days of free float. Node 5 is plotted at the end of the second day; node 7 already has been plotted at the end of the seventh day. The duration of 5-7, two days, added to its earliest start of three days, gives five days to plot between the end of activity 5-7 and node 7. This is the free float and is so indicated by means of a wavy line. Thus in the field, the superintendent can tell in a glance at the progress schedule, without having to read a computation table or computer printout, which activities need his immediate attention and which can be prolonged and for how long. This assistance is invaluable to him in his day to day allocation of manpower and equipment and ordering delivery of materials.

Dummy activities do not have to be shown on a calendar-time phased schedule but their indication can be helpful to one who is familiar with CPM. If shown on the schedule, those dummies with no free float would appear as vertical broken arrows. Where a dummy such as 19-22 figured to have some free float but had its free float struck out on form CPM 1, it would appear on the schedule as a slanted broken line.

To record physical performance periodically as the project progresses presents no new techniques to the project superintendent. The simplest method of accomplishing a task usually is the best one. The physical progress accomplished for each activity during any reporting period can be shown by overlaying a proportion of the arrow with red pencil. This is shown in Figure 7-2 as a superimposed wavy line. Transpired working days can easily be indicated by black pencil shading across the schedule from day to day. In Figure 7-2 it can be easily seen, as of a reporting period at the end of ten working days, that job layout, excavation, and make up of forms all are completed and that the rebars have been delivered to the job site. However, due to bad weather, only half of the forms have been erected and, as of now, it would require another five days to complete erection of forms. The installation of rebars now can be determined to require three days instead of two. All in all, as of the end of ten working days, it appears that the project is two days behind schedule. One cannot always be sure that this is so. In this case it is obvious because rebars, activity 10-12, had one day of free float. Its start is now delayed until the 15th day and its completion delayed until the 18th day. Formwork, activity 7-9, now has its completion delayed until the 15th day, followed by activity 9-12 of three days' duration, and completing the 18th day. The project now has two parallel critical paths until then. Should it have been determined that on the tenth day rebars, activity 10-12, would have required four days to perform rather than two, the critical path would have changed from 1-3, 3-7, 7-9, 9-12 to 1-10, 10-12, allowing one day of total float on 7-9 or 9-12. The superintendent now would need to change the direction of his concentration from activities no longer critical to those that have become critical.

To make this determination we need to update form CPM 1 on more complicated projects (see Figure 7-3). Updating will determine any change in the critical path, the concurrence of two or more parallel critical paths, and a new total project duration. It is questionable whether the project progress schedule itself needs to be updated at this point. A more intelligent decision can be made after CPM 1 is updated.

The mechanics of updating are demonstrated in Figure 7-3. Node numbers and activity codes remain the same. Durations must be altered to reflect the new conditions. Activities A, B, C, D, and E, having all been completed, now have zero durations. Activity F still requires five days to complete, so its duration is changed from six to five, and activity G is changed from two to three. All other durations remain the same at this first updating. The earliest starts and finishes as well as the latest starts and finishes for all activities that have been completed are now entered as ten days, the end of the day of the report. The remaining starts and finishes, early and late, now are recomputed with the technique described in Chapter 6.

It is now a 63-day job instead of a 61-day job. If nothing further is changed, the project would be completed two days late. Project management is confronted with the problem of returning the project to its scheduled completion time. How to do this at the least additional expense to the contractor will be discussed in a later chapter.

Updating a progress schedule may be required for several reasons. In the words of Robert Burns: "The best laid schemes of mice and men aft gang a-gley." Failure of timely delivery of materials, failure to secure approval of shop drawings, failure to secure adequate manpower,

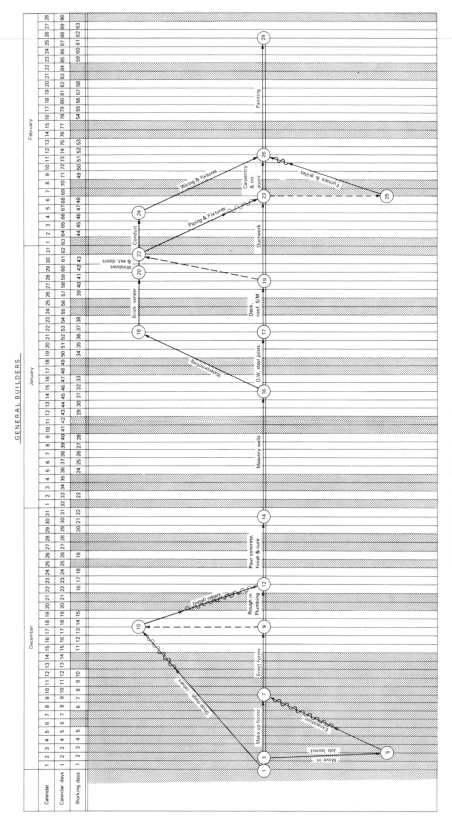

FIGURE 7-2 Small Warehouse Construction Progress Schedule.

Calculations for Float CPM 1

Job : Small Warehouse

Node No.		Act. Code	Description	Duration	Earliest		Latest		Float	
i	j				Start	Finish	Start	Finish	Total	Free
1	3	B	Move in	0	10	10	10	10	0	
1	10	E	Shop drawings - rebars	0	10	10	10	10	0	
3	5	A	Job layout	0	10	10	10	10	0	
3	7	D	Make up forms	0	10	10	10	10	0	
5	7	C	Excavation	0	10	10	10	10	0	
7	9	F	Erect forms	5	10	15	10	15	0	
9	10		Dummy	0	15	15	15	15	0	
9	12	H	Rough-in plumbing	3	15	18	15	18	0	
10	12	G	Install rebars	3	15	18	15	18	0	
12	14	I	Pour concrete, finish & cure	5	18	23	18	23	0	
14	16	J	Masonry walls	10	23	33	23	33	0	
16	17	K	O.W. steel joists	5	33	38	33	38	0	
16	18	L	Waterproofing	5	33	38	34	39	1	0
17	19	M	Deck, roof, & sheet metal	4	38	42	38	42	0	
18	20	N	Brick veneer	5	38	43	39	44	1	0
19	22		Dummy	0	42	42	46	46	4	
19	23	O	Ductwork	8	42	50	42	50	0	
20	22	P	Windows & ext. doors	2	43	45	44	46	1	0
22	23	Q	Piping & fixtures	3	45	48	47	50	2	2
22	24	R	Elect. conduit	3	45	48	46	49	1	0
23	25		Dummy	0	50	50	51	51	1	0
23	26	S	Carpentry & int. doors	3	50	53	50	53	0	
24	26	U	Wiring & fixtures	4	48	52	49	53	1	1
25	26	T	Furnace & grilles	2	50	52	51	53	1	1
26	28	V	Painting	10	53	63	53	63	0	

FIGURE 7-3

as well as unforeseen inclement weather—all would require updating the progress schedule, regardless of the type of schedule used. Critical path method cannot prevent any of the breakdowns cited when the failure is beyond the control of those responsible for performance. However, it can prevent some of the failures caused by error in timing or decisions. Nevertheless, even without serious failures along the line, good management requires that the schedule be updated periodically. There is no hard and fast rule that can be stated for the updating period. Jobs that are scheduled to take a year or more to build should have their schedules updated at least once per month. Jobs of lesser durations should have their schedule updated at least as often so as to provide competent control of the project. Aside from its updating on account of a failure, a project schedule should be updated, perhaps, in quarters of the project's duration.

The success of updating lies in the fact that it gives the project superintendent the opportunity to reevaluate the remaining activity durations in the light of actual experience already developed on this particular project. This reevaluation is of prime importance in redetermining the project duration and in doing so it may signal trouble ahead and provide the time required to assess the difficulty and plan suitable measures to surmount it.

CHAPTER 8

COSTS
OF
OPERATIONS

In Figure 7-3 it can be noted that even though the project duration has been extended to two days beyond the contract-allowed time for completion, there has been no change in the float times or in the critical path of the remaining activities. This condition may or may not be usual. Very often a delay in the finish of one activity may cause a subsequent activity to enlarge its duration. Also, with or without elongation of duration of subsequent activities, the delay in the finish of an activity may very well change the composition of the critical path.

Now would be the time for an administrative decision.

Certain factors need to be considered. Every job a building contractor undertakes has direct and indirect costs. The direct cost consists of the sum of the costs of labor, materials, and subcontracts incorporated into the building. Indirect costs are twofold. Costs incurred in supporting the direct costs, such as supervision, trucks, first aid, and the like are known as job overhead. Expenses incurred in maintaining a business, such as rent, interest, paper, estimating, and maintaining key personnel on the payroll in between jobs, are known as cost of doing business. This cost of doing business can be incurred at the job site as well as at the home office. Indirect costs can amount to as much as 20 percent of the direct costs and even may become greater than that when the life of a project is prolonged. Then too, whether the contractor is operating on his own capital or borrowing construction money from a bank, interest on the money continues for the additional time consumed on the job. There is also need to consider the time-delay damages clause of the contract. And last but not least are the additional expenses to a subsequent project when key men and equipment are required to remain on a job after another job needs their services for commencement of work.

On the other hand, it would take the outlay of additional money to speed up the remaining work to be done. All activities had been estimated at and the contract let for their most economical cost. The most economical cost, within the policies of this company, were based on performance at most efficient use of equipment and size of crew. A change in duration of any activity (if, indeed, a change can be effected) would cause a greater cost for that activity. This greater cost would vary from one activity to another.

Speeding up construction of an activity (it may be referred to as either "compression" or, to borrow a term from the military, "crashing") requires either going into overtime and/or weekend work, or double shifting, or acquiring additional equipment. Any of these would increase the cost of the activity. Another way to crash an activity would be to employ more men. This creates inefficiency in crew size and again increases the cost of the activity.

Even lengthening the duration of an activity may sometimes create a greater cost if that lengthening is accomplished by reduction of manpower, which reduces the efficiency of the crew.

In order to best solve the problem of being two days behind schedule, management must make a comparison of the additional expense of indirect costs and damages against the additional costs of crashing the project. The estimator now must develop sufficient information for management to make that decision. The information needed is a comparison of time required to perform certain activities at economical costs with the time these activities could be performed at increased costs. This is shown in Figure 8-1. As will be seen in a subsequent chapter, management would decide to crash certain activities in order to bring the project duration back into contract time at the least additional cost.

Time and Cost Comparison

Act.	Description	Op.	Normal				Crash				Difference		
			Dur.	Cost	ES	EF	Dur.	Cost	ES	EF	Dur.	Cost	Slope
B	Move in	1	1	(in O.H.)			1	(in O.H.)			0	—	—
E	Shop dwgs. - rebars	2	10	none			10	none			0	—	—
D	Make up forms	3	6	2,880			4	3,174			2	200	100
A	Job layout	4	1	90			1	90			0	—	—
C	Excavation	5	3	500			3	500			0	—	—
F	Erect forms	6	6	1,320			4	1,436			2	210	105
G	Install rebars	7	2	880			1	940			1	60	60
H	Rough-in plumbing	8	3	1,385			3	1,385			0	—	—
I	Pour conc. & cure	9	5	4,066			5	4,066			0	—	—
J	Masonry walls	10	10	7,615			7	7,735			3	120	40
K	O.W. steel joists	11	5	4,200			3	4,440			2	240	120
L	Waterproofing	12	5	738			3	778			2	40	20
M	Deck, roof & s.m.	13	4	5,478			2	5,598			2	120	60
N	Brick veneer	14	5	3,808			3	3,938			2	130	65
O	Ductwork	15	8	1,600			4	1,880			4	280	70
P	Windows & ext. drs.	16	2	970			2	970			0	—	—
Q	Piping & fixtures	17	3	1,015			2	1,095			1	80	80
R	Conduit	18	3	650			3	650			0	—	—
S	Carpentry & int. drs.	19	3	870			2	980			1	110	110
T	Furnace & grilles	20	2	2,000			2	2,000			0	—	—
U	Wiring & fixtures	21	4	2,850			4	2,850			0	—	—
V	Painting	22	10	1,750			10	1,750			0	—	—

FIGURE 8-1

Arriving at this decision is not quite so simple as it might appear. Its mechanics will be discussed in detail in subsequent chapters.

To make the comparison of time and cost, each activity must be analyzed individually. It should be a joint analysis of the estimator who does the work of preparation, the superintendent who provides the estimator with the information, and the manager who makes the final decision.

Normal duration and costs are taken from Figures 1-2 and 6-1, which were based originally on company facilities for execution of the project. The superintendent studies various methods by which an activity can be crashed and the estimator determines the new cost.

Not all activities can be shortened in duration. Those activities that can may be shortened by several different methods. Still other activities that could be crashed on one job could not be crashed on another job. No matter what methods, nor how much money, nor how many men are put to the task, some activities could not be performed in less time. Also, for those activities that can be crashed, there must be a limit to how far their duration can be shortened. There comes a time in any operation when, because of space limitations, drying out, curing, or whatever, its duration cannot be shortened any further no matter what additional expense can be incurred.

Time and money form the key to costs of operations. Normal durations based on economical cost were discussed in detail in Chapter 5. Crash durations and costs now will be discussed briefly. Each activity in the project needs to be considered individually in the light of economical or normal production. Depending on the nature of the work, the crew, and the equipment, the superintendent will decide which method to use for shortening the duration of the activity. He also will decide the amount of reduction in efficiency. For example, a bricklayer who normally could lay 500 bricks in an eight-hour day could be expected to lay 600 bricks, rather than 625 bricks, in a ten-hour day. If his base pay is $6.00 per hour, the cost per brick laid in an eight-hour day would be 9.6 cents per brick, whereas the cost per brick laid in a ten-hour day would be 12.0 cents per brick. The accelerated difference in cost is brought about by the fact that the bricklayer would earn $12.00 per hour for the two hours of overtime per day, and his efficiency is reduced. Increasing the size of the crew or adding another crew still would cause a loss of efficiency. The estimator would take the information given him by the superintendent—to wit: "If I had three more men or another piece of equipment, I could do this activity in two days' less time"—and compute the new cost for the activity.

To demonstrate the technique let us refer to Figure 8-1. The normal duration and cost of activity D, make up forms, are six days and $2,880, of which $1,200 is estimated to be labor cost for a crew of four carpenters and two laborers. The superintendent tells the estimator that if he had six carpenters and three laborers working ten hours per day he could make up the forms in four days. The estimator finds that under these conditions, the cost would be $3,174. The same procedure applies to activity F, erection of forms.

Likewise, activity G, install rebars, has a normal duration of two days and an economical cost of $880, of which $240 is labor cost based on employing three rodbusters working eight-hour days. The superintendent advises the estimator that he could tie and install all the reinforcing steel in one day if he had five rodbusters working ten-hour days. Again the estimator would recalculate the cost of this activity and would find it to be $940.

In studying activity J, masonry walls, there are 5,305 SF of masonry wall to erect. Normal time is ten days and normal cost is $7,615, distributed about equally between labor and material. The labor cost and time were based on employment of two crews working ten days. The superintendent decides that by the addition of another crew the three crews could do the work in seven days, accounting for some loss in efficiency through crowding, some loss in spreading supervision over three crews, and some expense in additional equipment to serve the third crew. The estimator now figures that the cost of masonry at crash performance would be $7,735. The saving in time would be three days, the additional cost would be $120, and the slope would be $40 per day. Other activities that can be crashed would follow the pattern of formwork and masonry, with the exception of those activities that are let to subcontractors. For these activities, the subcontractor would have to determine the degree of crash possible and the increased cost and so advise the estimator.

The determination of crashability, extent of shortening, and increased cost therefore should be determined for each and every activity in the job prior to start of construction or, if that is not possible, as soon as possible after the start. It is never too late to make this time and cost comparison, but if one waits until the job is in trouble, the chances are that there may be panic and good money would be wasted in attempts to speed up all the remaining activities when only one or two would be efficacious.

When crash durations and costs have been determined, the differences are simple to compute. To arrive at the amount of shortening that can be achieved, subtract the crash duration from the normal duration. To arrive at the difference in cost, subtract the normal cost from the crash cost. To arrive at the slope, divide the difference in cost by the difference in duration.

This process assumes a straight line slope of cost per day in crashing. This is not quite true, especially where the difference in duration is two days or more. Technically, studies have shown that time-cost curves are not straight lines and that shortening one operation induces a change in cost in other operations. These time-cost criteria are highly technical and complex. The purpose of this book is not to go into technicalities but to provide management with effective day to day tools that can be used reasonably well. It has been demonstrated that a straight line slope of cost per day would not deviate from an actual time cost curve to significant extent and can be used effectively in the construction field.

All the information necessary for crashing has now been developed, but not in readily adaptable form. Because crashing an operation may cause other activities to become critical and may itself become non-critical, it would be folly to spend money on any operation beyond the point where it would have become noncritical. The procedure for crashing is complicated, but by the use of forms CPM 2, CPM 3, and CPM 4, the procedure can become purely mechanical. In order to use these forms, we must rearrange the information gathered to this point and introduce some new terms and thoughts.

CHAPTER 9

GETTING READY
TO
CRASH

Up to this stage, arrow network diagrams and determination of floats have been used in the text. The arrow, bounded by starting and completion nodes, was thought of as an activity of work with the implication of duration ascribed to the arrow. Float was thought of in terms of slack time or leeway, or allowable delay in the starting of an activity, or in the lengthening of its duration.

All of this is well and good. It has forced management into a better understanding of the project. In actual practice, this is about as far as the construction industry has gone in the use of CPM, either by manual approach or by use of computer. But this is only a half measure. To derive the greatest advantage from CPM, it needs to be used to bring a project out of trouble and onto schedule again. Unfortunately, the arrow method of CPM does not easily lend itself to this stage. Fortunately, the circle notation does.

The logic is the same for both methods of representation. The difference lies in the fact that the circle has now become an operation of work. In the mental transition from arrow to circle diagramming, it would be helpful to change the identification of an item of work from an activity to an operation. Activities were drafted as arrows; operations are drafted as circles. There are no nodes in the circle notation. The logic of CPM is developed by the construction of connecting lines between circles. Then, too, activities were identified by letters of the alphabet, operations by numbers.

Figures 9-1a and 9-1b show a transition from arrow network diagramming to a circle notation network using the first nine activities or operations in the small warehouse job. The alphabetically lettered arrows are cross-referenced to operation numbers. They are both tied in to the item of work involved in Figure 8-1. The circle notation is read just the same as the arrow diagram. The circle notation reads that immediately upon the completion of operation 1, move in, operations 3 and 4, make up forms and job layout, can begin. The arrow diagram reads that activities A and D, job layout and make up forms, cannot begin until activity B, move in, is completed. It is interesting to note that there are no dummy operations required in the circle notation. Dummy activity 9-10 is not needed to show the logic in circle notation. It was used to show that activity G, install rebars, could not start until both activities E and F—shop drawings, approval, order and deliver for rebars and erection of forms—were completed. The same logic is expressed in circle notation simply by connecting the circle of operation 7 to both the circles of operations 2 and 6.

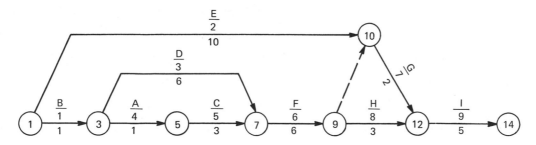

FIGURE 9-1a Arrow diagram—nine activities.

In arrow diagramming, the project network had to start with only one node and had to end with only one node. In circle notation, the project network may begin with as many circles as there are operations that can start the job. Also, if more than one operation can complete a

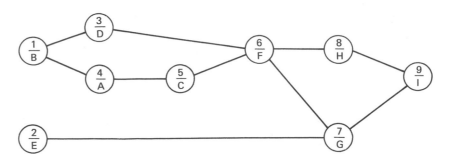

FIGURE 9-1b   Circle notation—eight operations.

project simultaneously, there could be as many circle operations as need be. However, in order that subsequent calculations may be made with circle notations, although they are not absolutely necessary, it would be permissible to introduce two circles of zero duration, one to read "Start" and one to read "End, " thus:

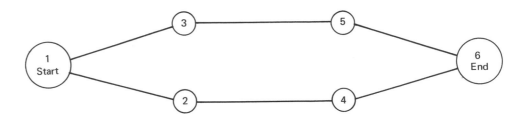

Another method of transition to a circle notation network could be made directly from a good bar chart (see Figure 9-2). In this case, the bar chart was made from the arrow diagram, Figure 6-8, and the cal-culations for earliest starts and finishes were made from Figure 6-3. A circle representing each item of work was made above each bar, and a number was placed inside the circle to identify the operation.   Then lines connecting the circle were drawn to indicate the logic already developed for the arrow diagram. Due to the fact that the bars are drawn to time scale, the circled numbers may not always be progressively forward, and some connecting lines may cross. Although crossing over of con-necting lines cannot be eliminated entirely, some can be eliminated by rearranging the circled numbers. Also due to time scaling of the bars, some connecting lines may fall too close together for clarity.   And, because for the purposes of crashing the estimator has to work with the connecting lines, it is advisable to show the circle notation on a separate work sheet, by itself and without regard to scaling of time, which from now on will be considered elsewhere. It is sufficient for now just to show the proper relationship of operations with respect to precedence. Refer to Figure 9-3 for a more workable circle notation network.

In our work with the arrow diagram, the term _float_ was defined. In work with the circle notation network, the term _lag_ appears and requires definition. Lag is the leeway between the earliest finish of a preceding operation and the earliest start of a following operation.

CPM logic may require that two concurrent operations be completed before a third operation can begin. When the two concurrent operations are of different durations, one may be critical, the other definitely is not. Therefore, the noncritical operation would have an earlier finish than the critical operation. But because the third operation cannot start until the completion of the critical operation, there is a time lag between the finish of the noncritical operation and the start of the third operation. This time lag is one of the determinants of how far the duration of the critical activity can be shortened and still reduce the duration of the project. When a critical operation has been shortened to the limit of the lag, the net effect is to make the noncritical activity become critical itself. Any shortening of the first critical operation beyond that lag would have the effect only of introducing a new lag between the finish of the first operation and the start of the third operation. In order to tabulate the lags existing in a project, there first must be a determination of the earliest starts and finishes of all operations in the project.

Although it would be possible to obtain the early start times and the early finish times from form CPM 1, Figure 6-3, completing the early starts and finishes schedule as shown in Figure 9-4 would facilitate the procedure for reference purposes. Another advantage in completing the normal group at this time lies in becoming more familiar with the circle notation network.

Operation numbers and durations are taken directly from Figure 9-2. Early starts and finishes are calculated from observation of the circle network, Figure 9-3, and application of the logic of CPM. Operations 1 and 2 start the project, concurrently, so their earliest start is entered as 0. Adding the duration of each operation to its ES gives its earliest

Job : Small Warehouse

Progress Schedule

Working Days

No.	Description	Dur.
1.	Move in	1
2.	Shop dwgs. – rebars	10
3.	Make up forms	6
4.	Job layout	1
5.	Excavation	3
6.	Erect forms	6
7.	Install rebars	2
8.	Rough-in plumbing	3
9.	Pour conc. & cure	5
10.	Masonry walls	10
11.	O.W. steel joists	5
12.	Waterproofing	5
13.	Deck, roof & s.m.	4
14.	Brick veneer	5
15.	Ductwork	8
16.	Windows & ext. drs.	2
17.	Piping & fixtures	3
18.	Conduit	3
19.	Carpentry & int. drs.	3
20.	Furnace & grilles	2
21.	Wiring & fixtures	4
22.	Painting	10

FIGURE 9-2

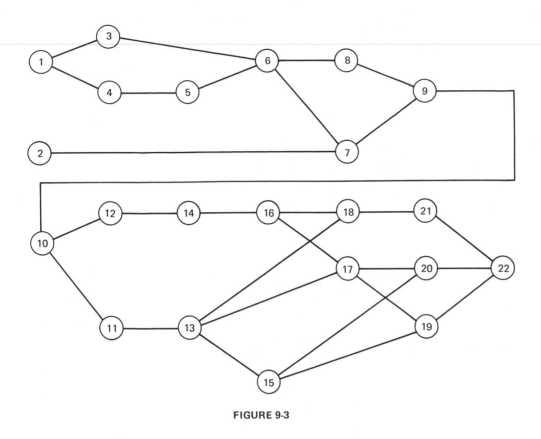

FIGURE 9-3

finish, 1 and 10, respectively. By looking at the circle network, one can see that op 3 and op 4 can start immediately upon the completion of op 1, whose EF is one. So, the ES for op 3 and op 4 is entered as one. Adding the six-day duration of op 3 to its ES of one gives its EF of seven.

Op 6 cannot start until both op 3 and op 5 have been completed. Op 3 has an EF of seven days and op 5 has an EF of five days. So the latter of the two possibilities must be used for the ES of op 6. Adding the duration of six days to the ES of op 6 of seven days gives its EF of 13 days. Because op 7 is dependent on both op 6 and op 2 for its ES, once again the latter of the two possibilities must be used for the ES of op 7. Completing the table, we arrive at a 61-day project duration, just as we did with the arrow network.

At this time, it would be advisable for the estimator to complete the second group of the table, all-crash. The same process of calculating ES and EF is followed through to arrive at a 49-day project duration. This would be the absolute minimum project duration, even though some operations were not crashed to their limit. Crashing all operations to their limit could not shorten the project duration any further and would result in an exercise of wasting money.

Time and Cost Comparison

Act.	Description	Op.	Normal				Crash				Difference		
			Dur.	Cost	ES	EF	Dur.	Cost	ES	EF	Dur.	Cost	Slope
B	Move in	1	1	(in O.H.)	0	1	1	(in O.H.)	0	1	0	—	—
E	Shop dwgs. – rebars	2	10	none	0	10	10	none	0	10	0	—	—
D	Make up forms	3	6	2,880	1	7	4	3,174	1	5	2	200	100
A	Job layout	4	1	90	1	2	1	90	1	2	0	—	—
C	Excavation	5	3	500	2	5	3	500	2	5	0	—	—
F	Erect forms	6	6	1,320	7	13	4	1,436	5	9	2	210	105
G	Install rebars	7	2	880	13	15	1	940	10	11	1	60	60
H	Rough-in plumbing	8	3	1,385	13	16	3	1,385	9	12	0	—	—
I	Pour conc. & cure	9	5	4,066	16	21	5	4,066	12	17	0	—	—
J	Masonry walls	10	10	7,615	21	31	7	7,735	17	24	3	120	40
K	O.W. steel joists	11	5	4,200	31	36	3	4,440	24	27	2	240	120
L	Waterproofing	12	5	738	31	36	3	778	24	27	2	40	20
M	Deck, roof & s.m.	13	4	5,478	36	40	2	5,598	27	29	2	120	60
N	Brick veneer	14	5	3,808	36	41	3	3,938	27	30	2	130	65
O	Ductwork	15	8	1,600	40	48	4	1,880	29	33	4	280	70
P	Windows & ext. drs.	16	2	970	41	43	2	970	30	32	0	—	—
Q	Piping & fixtures	17	3	1,015	43	46	2	1,095	32	34	1	80	80
R	Conduit	18	3	650	43	46	3	650	32	35	0	—	—
S	Carpentry & int. drs.	19	3	870	48	51	2	980	34	36	1	110	110
T	Furnace & grilles	20	2	2,000	48	50	2	2,000	34	36	0	—	—
U	Wiring & fixtures	21	4	2,850	46	50	4	2,850	35	39	0	—	—
V	Painting	22	10	1,750	51	61	10	1,750	39	49	0	—	—

FIGURE 9-4

One more observation should be made. Because op 6 cannot start until the seventh day and op 5 finishes at the end of the fifth day, there is a lag in time between the EF of op 5 and the ES of op 6. This lag is numerically equal to the free float ascribed to excavation.

Lag has already been defined as the amount of time, or the leeway, between the EF of a preceding operation and the ES of an immediately following operation. Following operations will now be referred to as post ops. By formula, lag can be expressed:

$$lag = ES \text{ (of post-op)} - E7 \text{ (of pre-op)}$$

The first step in crashing a job is the determination of lags. On the circle notation network, Figure 9-3, lag is represented as a connecting line between two operations without regard to scale. Because operations comprising the critical path must follow each other <u>immediately</u>, the lag between critical operations must, of necessity, be zero. That is not to say that all zero lags connect critical operations. There may be some connecting lines with zero lag that are not between critical operations. But there must be at least one chain (there could be two or more) of connecting lines of zero lag from start to finish of the project. To determine easily the lags of all connecting lines, form CPM 2 has been devised; see Figure 9-5. In preparing form CPM 2, every connecting line shown in the circle notation network must be represented by its pre-op and post-op. The post-operation cannot be listed in arithmetic order, but the pre-operations can and should be listed in increasing numerical order. Refer to Figure 9-4 for the normal ES and EF. Appropriate entries for post-ES and pre-EF then are entered in Figure 9-5. Then, by subtraction, the lags of each connecting line are easily obtained.

The second step is to transfer all zero lags back to the circle notation network, as shown in Figure 9-6a, by doubling all those connecting lines of zero lag. It now can be seen that there is but one chain of doubled connecting lines extending all the way through the project. That chain consists of lines connecting operations 1, 3, 6, 8, 9, 10, 11, 13, 15, 19, and 22. This is the identical critical path developed in earlier methods. This critical path is made clearer by the use of triple connecting lines, as shown in Figure 9-6b.

The third step is to set up form CPM 3, Figure 9-7. This is a graph showing the chain reaction effect on other operations when a specific operation is either shortened or lengthened. Across the top of the form are listed the following operations, or post-operations. Down the side are listed the preceding operations. A diagonal line is drawn as shown in Figure 9-7. Now all connecting lines having zero lag are indicated in the proper square by a small " o ". This can be obtained more readily from form CPM 2, Figure 9-5. All zero lags are entered by their pre-op and post-op numbers vertically down the first column headed lag. The first zero lag found is on the row of pre-op 1 and post-op 3, so a small circle is drawn on form CPM 3 on the row of pre-op 1 in the column of

Determination of Lags

CPM 2

Cycle					1		2		3		4		5		6		7	
Operation to change																		
Network limitation																		
Time changed																		
Pre-op	Post-op	Pre-EF	Post-ES	Lag		Lag		Lag		Lag		Lag		Lag		Lag		Lag
1	3	1	1	0														
1	4	1	1	0														
2	7	10	13	3														
3	6	7	7	0														
4	5	2	2	0														
5	6	5	7	2														
6	7	13	13	0														
6	8	13	13	0														
7	9	15	16	1														
8	9	16	16	0														
9	10	21	21	0														
10	11	31	31	0														
10	12	31	31	0														
11	13	36	36	0														
12	14	36	36	0														
13	15	40	40	0														
13	17	40	43	3														
13	18	40	43	3														
14	16	41	41	0														
15	19	48	48	0														
15	20	48	48	0														
16	17	43	43	0														
16	18	43	43	0														
17	19	46	48	2														
17	20	46	48	2														
18	21	46	46	0														
19	22	51	51	0														
20	22	50	51	1														
21	22	50	51	1														

FIGURE 9-5

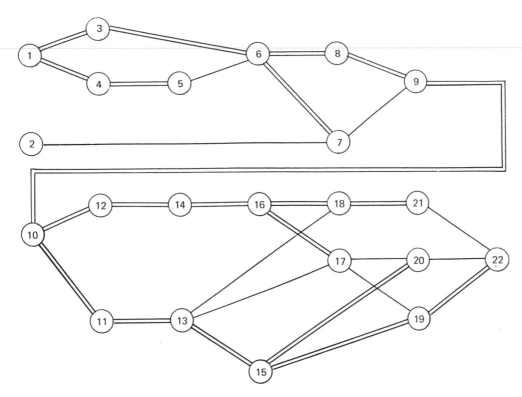

FIGURE 9-6a

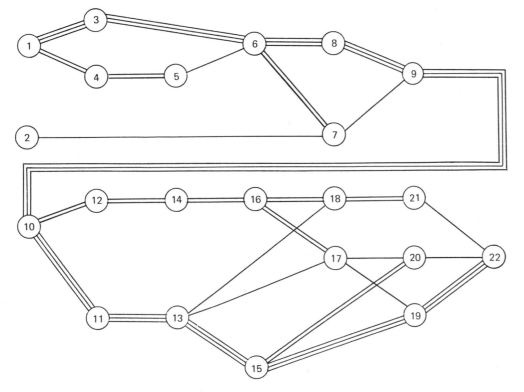

FIGURE 9-6b

post-op 3, and so on until every zero lag has been indicated on form **CPM 3.** The last zero lag found in Figure 9-5 is in the row of pre-op 19, post-op 22, so the last small circle entered in Figure 9-7 would be in the row of pre-op 19 under the column of post-op 22.

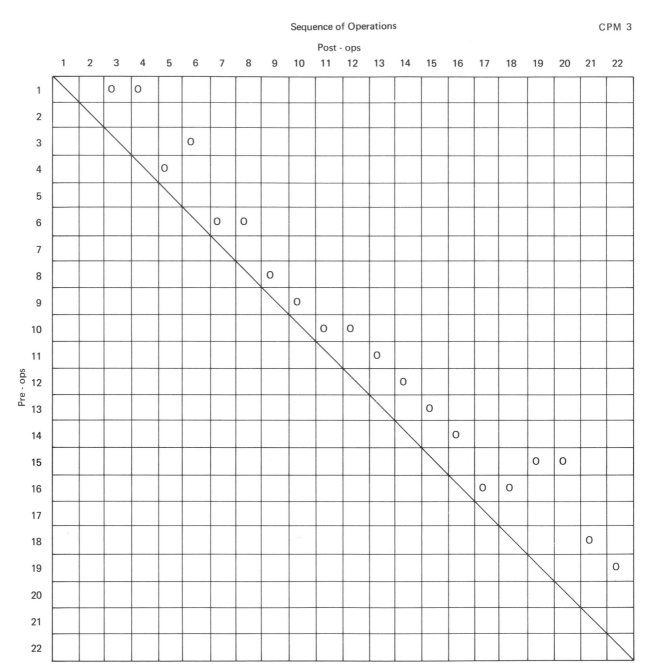

FIGURE 9-7

The significance of these circles is that in every column of post-ops where they appear, that operation must follow immediately, without any delay in time, the preceding operation of the row in which the circle appears. For example, there is a circle in column 10, operation masonry walls. The masonry work must be started immediately upon completion of operation 9, curing concrete, because the circle in column 10 also lies in the row of pre-op 9.

In the logic of CPM—that there must be at least one critical path, one chain of operations where each succeeding operation must begin immediately upon the completion of its preceding operation—there must be at least one circle in every column in form CPM 3 (every column except the first one, of course). There is no post-operation 1, because it starts the job, and there is no pre-operation 22, because it completes the job.

Changing the duration of an operation either by shortening or lengthening may or may not have a material effect on the project duration or other operations within the project. Shortening the last operation obviously shortens the project duration without affecting the earliest starts or finishes of any other operation. Shortening or lengthening the duration of the first operation will advance or delay the earliest starts and finishes of every operation in the project. Other than the first and last operations, any change in duration will affect the earliest starts and finishes of one or more operations but not all of them; and management must know which ones. This can be worked out laboriously, directly from the circle notation network. However, this same information can be gleaned from form CPM 3 mechanically, if a system of small x's is used to indicate the chain of reactions brought about by the change in duration of any specific operation. For example, if operation 14, brick veneer, were to be completed two days earlier than scheduled, then the following operation, 16, windows and exterior doors, would have its ES and EF advanced by two days. But by advancing the EF of op 16 by two days, the ES of op 17 and op 18 would also advance by two days. Op 18 is immediately followed by op 21 and so it too would have its ES and EF advanced by two days. Op 17 is followed by op 19 and op 20; op 21 is followed by op 22. The net effect of finishing noncritical operation 14 two days earlier than scheduled would be to reduce the lag between operations 13 and 17, as well as the lag between operations 13 and 18 from three days to one day, and increase the lag between operations 17 and 19 as well as between operations 17 and 20 from two days to four. The lag between operations 21 and 22 also would be increased by one day. If any expense had been incurred in reducing the duration of operation 14, it would have been wasted money insofar as this small warehouse is concerned.

The sequence of operations cited can readily be seen in Figure 9-8 simply if one reads across the row of pre-op 14 and notes the circle in post-op column 16, and the x's in post-op columns 17, 18, and 21.

Form CPM 3 shows, then, that coordinates containing circles signify that there is no lag between the start of the post-op and the finish of the pre-op. Across the row of any pre-op, the operations in those columns containing circles and x's will have their EF and ES changed in accordance with the change in the pre-op of the row under consideration.

The mechanics of placing x's in the proper squares start with the last circle on form CPM 3, which occurs on row 19, column 22. Moving to the left on row 19 until the diagonal line is met in column 19, thence moving up column 19 until a circle is found on row 15, thence moving to the right on row 15, the project manager places an x in column 22. Now the next circle is found on row 18, in column 21. Moving to the left on row 18 until the diagonal line is met in column 18 and going up column 18 until a circle is found on row 16, then moving to the right, he places an x on row 16 in the columns containing the circle on row 18. The next circle is found on row 16. Moving to the left on row 16 until the diagonal line is met in column 16, and going up column 16 until a circle is found on row 14, and moving to the right on row 14, he places x's in the columns where circles and x's occur on row 16, which are at columns 17, 18, and 21. This procedure is followed through to completion in Figure 9-8. For any operation for which a change in duration is being considered, Figure 9-8 will show all other succeeding operations whose early starts and finishes will be affected by the change.

The fourth step in crashing consists of selecting operations to be crashed. To help make this selection, form CPM 4, Operation Selection Worksheet, is used (see Figure 9-9). This form is divided into two parts. The upper part of the form contains a five-part column headed Critical Paths; the lower part of the form contains a triple column for listing all the operations that can be physically crashed within themselves and showing their operation numbers, their slopes, and the amount of time crashable. Both parts of the form follow with several cycles of crashing.

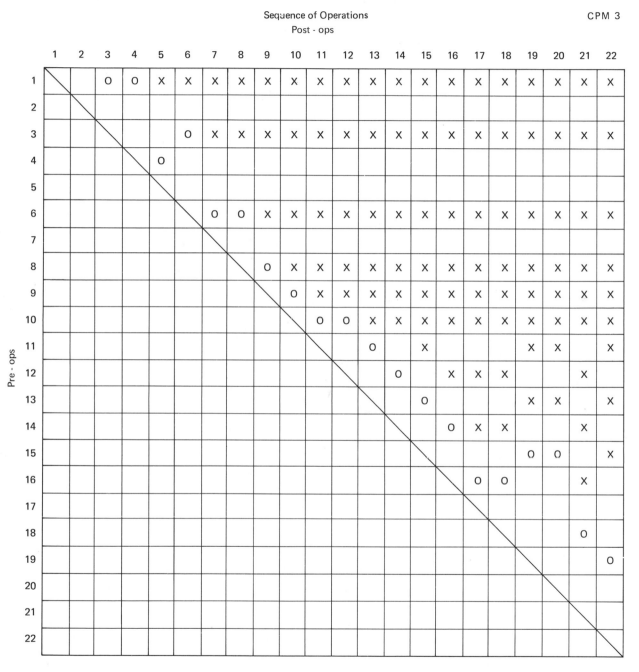

FIGURE 9-8

Operation Selection Worksheet CPM 4

Critical Paths					Cycle 1	Cycle 2	Cycle 3	Cycle 4	Cycle 5	Cycle 6	Cycle 7
1											
3											
6											
8											
9											
10											
11											
13											
15											
19											
22											

Crash-ops										
Op.	Slope	Time								
12	20	2								
10	40	3								
7	60	1								
13	60	2								
14	65	2								
15	70	4								
17	80	1								
3	100	2								
6	105	2								
19	110	1								
11	120	2								
	slope									
	time									
	add'l cost									

FIGURE 9-9

To use this form effectively, under the upper heading Critical Paths, list in numerical order those operations that are critical. Under the lower heading of Crash-Op, list all those operations that could be crashed in increasing order of slope, slope being the additional cost per day for crashing an operation.

The information for the upper listing is taken from the triple connecting lines in the circle notation, Figure 9-6b, and the information for the lower listing is taken from the time and cost comparison, Figure 9-4. To complete the initial setting up of the operation selection worksheet, strike from the upper part those critical operations, such as 1, 8, 9, and 22, which do not appear in the lower part. They cannot be crashed within themselves and are not to be considered.

This completes the preparatory work and we now are ready to start crashing this project.

CHAPTER 10

STAGE IV
OF
CPM SCHEDULING

THE CRASHING PROCESS

Before going into the mechanics of crashing a project, it would be wise to review the reasons why a builder would want to shorten the duration of a project.

First of all, many contract specifications require completion of a proposed project by a specific calendar date or in a specific number of days, either calendar or working days. In addition, there is usually a time delay damages clause in the contract whereby the contractor agrees to a reduction in the contract price in amount equal to a specific charge per day for each day the work continues beyond the contract period. Having completed his proposed bid, the contractor must know whether he can complete the project on time. If he can not, he should increase the amount of his bid by the amount of the demurrage. If in doing so he feels that his competitors may beat him, then he should at least increase his proposed bid by an amount equal to the additional cost to him incurred in performing his contract within the specified time.

In the second instance, if after construction has started, a periodic progress report indicates that the project is falling behind schedule, it would be good to know what could be done to bring the project back on schedule at the least additional cost to the builder. In this respect, after spending some money in order to shorten the remaining project duration, the builder may find it cheaper to pay the demurrage after a certain point in shortening.

Rarely does an owner offer a bonus as an inducement to the contractor to complete a project ahead of a specified time. This type of inducement is usually in the form of a specific amount per day for every day saved on the job. Once again, there may come a point in the speedup of operations where it would cost more per day to crash than the amount per day offered by the bonus. The builder would want to stop crashing at that point.

With the exception of wartime activities, in which cost is no object, very seldom would a building job ever be crashed to its utmost limit. By the same token, very seldom is there a project that does not need to be crashed to some extent. It is good planning to be ready to meet any job emergency, no matter how remote, and in that sense the small ware-house job of this text will be crashed to its limit. In doing so, a record will be kept at the bottom of form CPM 4 showing the accumulated shortening and costs at each cycle.

To shorten a project duration, to crash a job, three factors must be taken into account in the selection of the operation to be crashed.

1. The operation must lie on the critical path.
2. The operation must be able to be crashed.
3. The operation must have the least slope.

Because the project duration is determined by the chain of critical oper-ations, and unless the operation to be crashed is critical, the project duration would not be shortened. Whenever a noncritical operation is crashed, the effect is twofold. First, the lag between it and its following operation would be increased. And second, the additional cost of crashing would be money thrown away.

The slope is the additional cost per day for shortening the duration of an operation. Common sense dictates that when presented with a choice of two or more operations to be crashed, the builder selects the one that costs the least.

Looking at Figure 9-9, we can see that the critical operations that can be crashed are 3, 6, 10, 11, 13, 15, and 19. The first operation listed in the lower part is 12. It has the least slope, but it is not critical. Skip it. The next operation is 10 with a slope of $40 per day. It is critical and it could be crashed as much as three days if the interrelationship of operations would allow that much. In shortening the duration of a critical operation, there always is the possibility that before it reaches its physical limit of crashing, some other noncritical operation has become critical on account of loss of lag. In such an instance, it would be useless to crash the operation by itself to its limit. However, operation 10 is selected for crashing the first cycle. This fact is indicated by entries made on form CPM 4, Figure 10-1. In the upper part, under Cycle 1 on the row of operation 10, is entered the number 10 placed in a circle. In the lower part of the form, still in the column of cycle 1, on the row of op 10, a small x is placed in a circle. On the line of slope under the x is placed the figure 40 circled, which is the cost per day for crashing operation 10.

Now form CPM 2, Figure 10-2, is entered. In cycle 1, opposite operation, is entered 10; opposite change is entered 3, the time that operation 10 could be crashed if the network would so allow. The network limit for crashing operation 10 now needs to be determined. The work done in preparing form CPM 3, Figure 9-8, is now brought into play. Looking at the row of pre-op 10, there can be found circles and x's under columns of post-ops 11, 12, 13, 14, 15, 16, 17, 18, 19, 20, 21, and 22. It already has been learned, in Chapter 9, that when the EF of an operation is advanced (by shortening its duration), all of the remaining operations, in sequence, have their ES and EF advanced by an equal amount. This information must be entered on form CPM 2, Figure 10-2, by means of horizontal arrows indicating the advance of the EF and ES of the operations involved. In placing the horizontal arrows, two columns of arrows are formed. The left-hand column indicates that the EF of the pre-op in that row is being advanced and the right-hand column indicates that the ES of the post-op in that row is being advanced. As will be seen in a later cycle, it can be deduced that where the ES of a

Operation Selection Worksheet   CPM 4

Critical Paths					Cycle 1	Cycle 2	Cycle 3	Cycle 4	Cycle 5	Cycle 6	Cycle 7
1											
3											
6											
8											
9											
10					⑩						
11											
13											
15											
19											
22											

Crash-ops											
Op.	Slope	Time									
12	20	2									
10	40	3	Ⓧ								
7	60	1									
13	60	2									
14	65	2									
15	70	4									
17	80	1									
3	100	2									
6	105	2									
19	110	1									
11	120	2									
	slope		㊵								
	time		x3								
	add'l cost		120								

FIGURE 10-1

post-operation is being advanced without a concomitant advance in the EF of a pre-operation, there evolves a reduction in lag. When more than one such situation occurs, the row containing the least amount of lag is the situation that limits the shortening of the operation attempted. The number of days of lag is the network interaction limit and is so entered in the appropriate cycle on form CPM 2.

Looking at Figure 10-2, we find that in cycle 1 of this project there does not occur any advance in ES of a post-operation without an advance in EF of a pre-operation. So the network limitation is none and the word none is entered as the network limitation. An examination of the circle notation network, Figure 9-6b, would substantiate this result. Because three days is the utmost that operation 10 can be physically crashed and there is no network limitation, the decision would be made to crash operation 10 by three days.

This decision is recorded on form CPM 4, Figure 10-1, in the following manner. In the column of cycle 1, opposite time (to be crashed) is entered the number 3 with the indication of the multiplier, so that the additional cost to the builder of crashing operation 10 by three days is $120. At this point, form CPM 2, Figure 10-2, must be reentered for computation of new lags brought about by this crashing. Cycle 1, having no network limit, changes no lags, so all previous lags are brought forward to be the remaining lags at the end of cycle 1. There is no need to bring forward any zero lags. With no complications presented by this action, cycle 1 of crashing is completed.

Before starting cycle 2, operation 10 must be struck from form CPM 4, Figure 10-3, because it was crashed all the way in cycle 1 and no longer can be considered. It is still an easy matter to select the operation for crashing in cycle 2 because there still is but one critical path. Looking at the lower part of Figure 10-3, notice that the next crashable operation in order of increasing slope is 7. In the upper part, operation 7 does not appear to be critical, so operation 13, which is the next operation in the lower part, is looked for in the upper part. It is found and the number 13 is placed opposite operation 13 under cycle 2. An x is placed opposite operation 13 in the lower half, and because no other operations need be considered for cycle 2, both entries as well as the slope of 60 are placed in circles to indicate the selection. In form

Determination of Lags   CPM 2

Cycle					1		2		3		4		5		6		7	
Operation to change					10 3													
Network limitation					none													
Time changed					3													
Pre-op	Post-op	Pre-EF	Post-ES	Lag		Lag		Lag		Lag		Lag		Lag		Lag		Lag
1	3	1	1	0														
1	4	1	1	0														
2	7	10	13	3		3												
3	6	7	7	0														
4	5	2	2	0														
5	6	5	7	2		2												
6	7	13	13	0														
6	8	13	13	0														
7	9	15	16	1		1												
8	9	16	16	0														
9	10	21	21	0														
10	11	31	31	0	←|←													
10	12	31	31	0	←|←													
11	13	36	36	0	←|←													
12	14	36	36	0	←|←													
13	15	40	40	0	←|←													
13	17	40	43	3	←|←	3												
13	18	40	43	3	←|←	3												
14	16	41	41	0	←|←													
15	19	48	48	0	←|←													
15	20	48	48	0	←|←													
16	17	43	43	0	←|←													
16	18	43	43	0	←|←													
17	19	46	48	2	←|←	2												
17	20	46	48	2	←|←	2												
18	21	46	46	0	←|←													
19	22	51	51	0	←|←													
20	22	50	51	1	←|←	1												
21	22	50	51	1	←|←	1												

FIGURE 10-2

CPM 2, Figure 10-4, under column of cycle 2, the operation 13 is indicated as well as the two days it could be crashed barring network limitation. Again to determine the network limitation, form CPM 3, Figure 9-8, is read. On the row of pre-op 13, circles and x's are found in the post-op columns of 15, 19, 20, and 22. Once again form CPM 2, Figure 10-4, is marked under column of cycle 2. Horizontal arrows are placed on the left-hand side in the rows where pre-ops 13, 15, 19, and 20 are found. There is no pre-op 22. Horizontal arrows also are placed on the right-hand side in the rows where post-ops 15, 19, 20, and 22 are found. There are three instances where the arrows indicate an advance in ES of a post-operation without advancing the EF of a pre-operation. They are the combinations of 17 and 19, 17 and 20, and 21 and 22. The lag of the first combination is two days; the lag of the third combination is one day. One day being the least, it is the network limitation to crashing operation 13 and is so indicated by placing the horizontal arrow in a circle and entering the number 1 opposite the network limitation and the time changed. The recomputed lags now are entered for the end of cycle 2. It is interesting to note that the EF of operation 13 is advanced without advancing the ES of post-operations 17 and 18. In these cases, the lag of three days at each is increased to four days. In the case of pre-op 21 and post-op 22, the one-day lag is reduced by the one day of time and changed to zero. All the worksheets will have to be updated before going into cycle 3 in order to reflect this new zero lag. However, before updating the worksheets, form CPM 4, Figure 10-3, needs to be completed for cycle 2 by the indication of time shortened and cost added. Also, in the lower part, opposite operation 13, the time needs to be changed from two days to one day, because it was not crashed all the way and remains available for a later cycle.

Before starting cycle 3, the worksheets must be updated on account of the new zero lag between the EF of pre-op 21 and the ES of post-op 22. The line connecting these two operations needs to be doubled on the circle notation network, as shown in Figure 10-5a. There now comes a new subchain of connecting lines of zero lag, a parallel critical path as it were, from operation 10 through 12, 14, 16, 18, 21, and 22. The lines connecting these operations now must be tripled, as shown in Figure

Operation Selection Worksheet CPM 4

Critical Paths					Cycle 1	Cycle 2	Cycle 3	Cycle 4	Cycle 5	Cycle 6	Cycle 7
1											
3											
6											
8											
9											
10					⑩						
11											
13						⑬					
15											
19											
22											

Crash-ops										
Op.	Slope	Time								
12	20	2								
10	40	3	Ⓧ							
7	60	1								
13	60	2 1		Ⓧ						
14	65	2								
15	70	4								
17	80	1								
3	100	2								
6	105	2								
19	110	1								
11	120	2								
	slope		㊵	60						
	time		x3 +	x1 = 4						
	add'l cost		120 +	60 = 180						

FIGURE 10-3

Determination of Lags   CPM 2

Cycle					1		2		3		4		5		6		7	
Operation to change					10	3	13	2										
Network limitation					none		1											
Time changed					3		1											
Pre-op	Post-op	Pre-EF	Post-ES	Lag		Lag		Lag		Lag		Lag		Lag		Lag		Lag
1	3	1	1	0														
1	4	1	1	0														
2	7	10	13	3		3		3										
3	6	7	7	0														
4	5	2	2	0														
5	6	5	7	2		2		2										
6	7	13	13	0														
6	8	13	13	0														
7	9	15	16	1		1		1										
8	9	16	16	0														
9	10	21	21	0														
10	11	31	31	0	←|←													
10	12	31	31	0	←|←													
11	13	36	36	0	←|←													
12	14	36	36	0	←|←													
13	15	40	40	0	←|←		←|←											
13	17	40	43	3	←|←	3	←|	4										
13	18	40	43	3	←|←	3	←|	4										
14	16	41	41	0	←|←													
15	19	48	48	0	←|←		←|←											
15	20	48	48	0	←|←		←|←											
16	17	43	43	0	←|←													
16	18	43	43	0	←|←													
17	19	46	48	2	←|←	2	|←	1										
17	20	46	48	2	←|←	2	|←	1										
18	21	46	46	0	←|←													
19	22	51	51	0	←|←		←|←											
20	22	50	51	1	←|←	1	←|←	1										
21	22	50	51	1	←|←	1	|⊖	0										

FIGURE 10-4

10-5b, to show that they have become critical and in parallel to operations 10, 11, 13, 15, 19, and 22.

Form CPM 3, the Sequence of Operations worksheet, needs to be updated, as shown in Figure 10-6. This step probably is the most difficult. First, a circle needs to be placed in the row of pre-op 21 under the column of post-op 22. Later on, in the crashing process, it will be demonstrated that new x's will need to be entered, first working toward the end of the project, then working toward the beginning. However, because operation 22 is the last operation in this project, updating can be performed only from this new circle toward the beginning of the project. Starting with this new circle, move to the left on row of pre-op 21 until the diagonal line is met in the column of post-op 21, move up this column until a circle is found on the row of pre-op 18, move to the right on this row and place x's in the columns where circles and x's occur on row of pre-op 21. Only one x is placed in row 18, and that is in column 22 because column 22 is the only column containing a circle or an x on row 21. But the process cannot stop here. A new x has been entered on row 18, column 22, so the updating must continue until no new x's are introduced. On row of pre-op 18, move to the left until the diagonal line is met in column of post-op 18, move up column 18 until a circle is found on the row of pre-op 16, move to the right on row 16 and place an x in column 22, above the x that had been introduced on row 18. Continue the process until the last new x is entered on row 12, column 22.

Form CPM 4 now can be updated, as shown in Figure 10-7. In the upper part, all of the new critical operations—beginning with operation 10 and ending with operation 22—must be listed in the second column of critical paths. Comparing these new critical operations with the crash operations listed in the lower part of the form, we can see that operations 16, 18, and 21 do not appear. They cannot be crashed and should be struck out of the upper part. Operation 10 already has been crashed all the way, so it too needs to be struck from further consideration.

The worksheets now have been updated and are ready for crashing cycle 3. But the selection of the operations or operation no longer is simple. There now are two possibilities for shortening the project duration. Either operation 3 or 6 alone would shorten the project duration because there is only the one critical path in this part of the network.

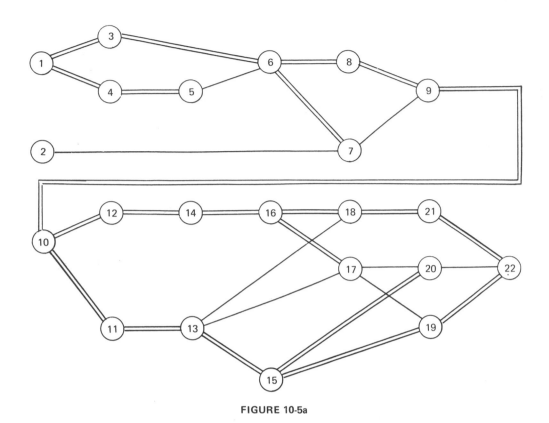

FIGURE 10-5a

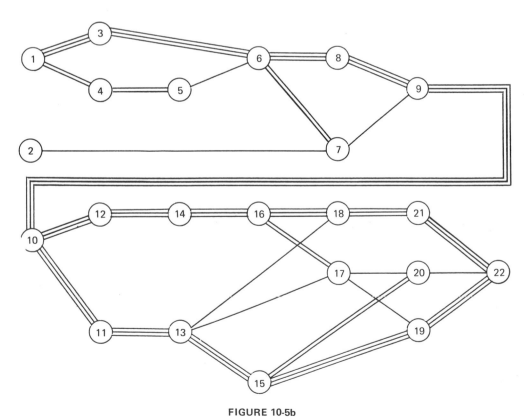

FIGURE 10-5b

Sequence of Operations CPM 3

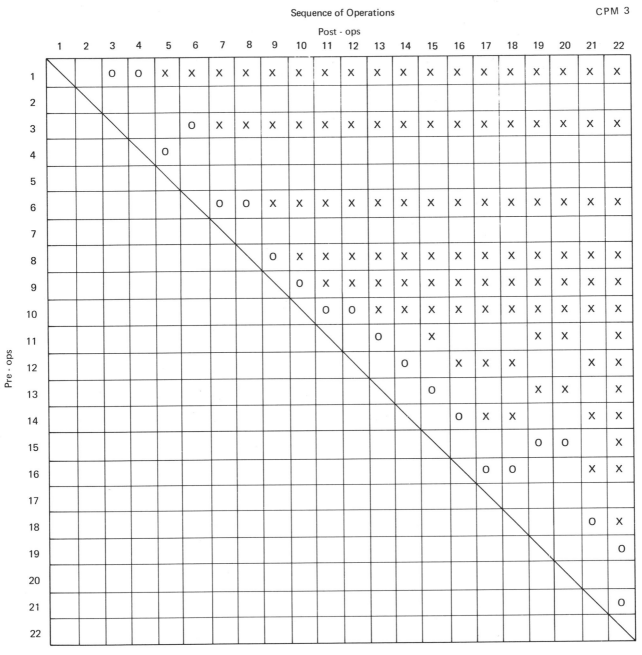

FIGURE 10-6

Because 3 is found before 6 in the lower part of the form, operation 3 is considered for crashing by placing the number 3 in the column of cycle 3. Opposite operation 3 in the lower part, an x is placed in column of cycle 3, and the slope of 100 is entered on the row of slopes. But this may not be the most economical operation to crash at this time. Certainly operation 12, which has now become critical, could be crashed at a cost of only $20 per day, so the number 12 is entered in column of cycle 3 opposite its row in the upper part. But crashing operation 12 alone, lying as it does on one of two parallel critical paths, cannot reduce the project duration, <u>unless a critical operation lying on the other critical path is crashed in tandem with it.</u> The operations found on the other path are 11, 13, 15, and 19. Looking down the list of crash-ops for the one with least slope, operation 13 is the first we see. So, in the column of cycle 3, immediately under 12 is placed 13, and in the lower half, x's are placed opposite 12 and 13, one below the other. The sum of the slopes represented by these 2 x's are 20 plus 60, or 80, which is entered at the bottom opposite the 100 that already had been entered. At this time, no further study of combinations of operations is warranted because the operation with the least slope in each of the two critical paths is under consideration.

It now appears that crashing operations 12 and 13 in tandem at a cost of $80 per day is more economical to the builder than crashing operation 3 alone at a daily cost of $100, and this fact is so indicated by placing circles around the operations and slopes selected for crashing cycle 3.

This choice is now entered under the column of cycle 3, on form CPM 2, as shown in Figure 10-8. Operation 12 can be crashed two days, but there is only one day left to crash in operation 13, so the desired shortening of the tandem is limited to one day. In analyzing the effect of shortening by one day, should it appear that the network limitation works out to be zero, it is indicative of a mistake made previously, or an error made in selection of the proper operations to be crashed. Referring to form CPM 3, Figure 10-6, for the sequence of operations affected by the shortening of the durations of operations 12 and 13, we find that each one needs to be considered individually. With respect to pre-op 12, operations 14, 16, 17, 18, 21, and 22 are affected. With respect to pre-op 13, operations 15, 19, 20, and 22 are affected. This information is entered on form CPM 2, Figure 10-8, by means of the horizontal arrows. Once again, because no post-op has its ES advanced without a concomitant advance of the EF of a pre-op, the network limitation is "none," and the time changed is one day. The lags for all connecting lines at the end of cycle 3 would be the same as those at the end of cycle

Operation Selection Worksheet   CPM 4

Critical Paths				Cycle 1	Cycle 2	Cycle 3	Cycle 4	Cycle 5	Cycle 6	Cycle 7
~~4~~										
3						3				
6										
~~8~~										
~~9~~										
~~10~~	~~10~~			⑩						
11										
	12					⑫				
13					⑬	⑬				
	14									
15										
	~~16~~									
	~~18~~									
19										
	~~21~~									
~~22~~	~~22~~									
Crash-ops										

Op.	Slope	Time								
12	20	2				ⓧ				
~~10~~	40	~~2~~	ⓧ							
7	60	1								
13	60	~~2~~ 1			ⓧ	ⓧ				
14	65	2								
15	70	4								
17	80	1								
3	100	2				X				
6	105	2								
19	110	1								
11	120	2								
	slope			④⓪	⑥⓪	100 ⑧⓪				
	time			$\underline{\times 3}$ +	$\underline{\times 1 = 4}$ +	$\underline{\times 1 = 5}$				
	add'l cost			120 +	60 = 180 +	80 = 260				

FIGURE 10-7

2. Form CPM 4, Figure 10-7, is now completed at the bottom to show that through cycle 3, the project duration can be shortened by as much as five days at a total cost of $260.

Before starting cycle 4, form CPM 4 needs to be updated, as shown in Figure 10-9. Cycle 3 shortened the durations of operations 12 and 13 by one day. The crashable time remaining to these two operations must be changed to reflect this. Operation 12 still has one day left, but operation 13 now has been crashed to its limit and must be struck from both parts of form CPM 4. Operation 3, which was considered and rejected in cycle 3, needs to be reconsidered as a possibility for cycle 4. Operation 12 has yet one day remaining and should be considered in tandem with operation 11, 15, or 19, whichever appears first in the lower part. It is operation 15 at a slope of 70 that when added to 20, the slope of operation 12, gives a sum of 90 for crashing the tandem. Obviously, it is more economical to crash operations 12 and 15 in tandem than to crash operation 3 alone. The process of determining the network limit and redetermining lags is repeated on form CPM 2, Figure 10-10. It is found on the two rows of pre-op 13—post-op 17 and pre-op 13—post-op 18 that the ES of post-ops 17 and 18 are advanced without advancing the EF of pre-op 13, that the lag in both cases is four days and therefore four days is the network limit for crashing operations 12 and 15. But there was only one day remaining to operation 12, so the decision is made to crash these operations by one day. In redetermining the lags, the four-day lags just discussed will be reduced to three. No other change in lags is indicated. When the record at the bottom of Figure 10-9 is completed, it can be seen that at the end of cycle 4, the project duration can be shortened by six days at a cost of $350.

Before cycle 5 is started, form CPM 4, Figure 10-11, needs to be updated to reflect the shortening of operations 12 and 15 in cycle 4. Operation 12 now has been crashed all the way and should be struck out in both parts of the form. Operation 15 needs to show that it has three days of crash left. Because cycle 4 did not introduce any new zero lags and so did not change the circle notation network, operation 3 will be reconsidered for crashing in cycle 5. Operation 14 now stands alone in the second critical path, so it needs to be considered in tandem with operation 11, 15, or 19. Cycle 4 determined that operation 15

Determination of Lags

CPM 2

					1		2		3		4		5		6		7	
Cycle					1		2		3		4		5		6		7	
Operation to change					10 3		13 2		12, 13 1									
Network limitation					none		1		none									
Time changed					3		1		1									
Pre-op	Post-op	Pre-EF	Post-ES	Lag		Lag		Lag		Lag		Lag		Lag		Lag		Lag
1	3	1	1	0														
1	4	1	1	0														
2	7	10	13	3		3		3		3								
3	6	7	7	0														
4	5	2	2	0														
5	6	5	7	2		2		2		2								
6	7	13	13	0														
6	8	13	13	0														
7	9	15	16	1		1		1		1								
8	9	16	16	0														
9	10	21	21	0														
10	11	31	31	0	←|←													
10	12	31	31	0	←|←													
11	13	36	36	0	←|←													
12	14	36	36	0	←|←				←|←								
13	15	40	40	0	←|←		←|←		←|←								
13	17	40	43	3	←|←	3	←|	4	←|←	4								
13	18	40	43	3	←|←	3	←|	4	←|←	4								
14	16	41	41	0	←|←				←|←								
15	19	48	48	0	←|←		←|←		←|←								
15	20	48	48	0	←|←		←|←		←|←								
16	17	43	43	0	←|←				←|←								
16	18	43	43	0	←|←				←|←								
17	19	46	48	2	←|←	2	|←	1	←|←	1								
17	20	46	48	2	←|←	2	|←	1	←|←	1								
18	21	46	46	0	←|←				←|←								
19	22	51	51	0	←|←		←|←		←|←								
20	22	50	51	1	←|←	1	←|←	1	←|←	1								
21	22	50	51	1	←|←	1	|⊕	0	←|←								

FIGURE 10-8

Operation Selection Worksheet   CPM 4

Critical Paths					Cycle 1	Cycle 2	Cycle 3	Cycle 4	Cycle 5	Cycle 6	Cycle 7
~~1~~											
3							3	3			
6											
~~8~~											
~~9~~											
~~10~~	~~10~~				⑩						
11											
	12						⑫	⑫			
~~13~~						⑬	⑬				
	14										
15								⑮			
	~~16~~										
	~~18~~										
19											
	~~21~~										
~~22~~	~~22~~										

Crash-ops											
Op.	Slope	Time									
12	20	~~2~~ 1					ⓧ	ⓧ			
~~10~~	40	~~3~~			ⓧ						
7	60	1									
~~13~~	60	~~2~~ 1				ⓧ	ⓧ				
14	65	2									
15	70	4						ⓧ			
17	80	1									
3	100	2					X	X			
6	105	2									
19	110	1									
11	120	2									
	slope				㊵	㊿ 60	100 ⑧⓪	100 ⑨⓪			
	time				x3 +	x1 = 4 +	x1 = 5 +	x1 = 6			
	add'l cost				120 +	60 = 180 +	80 = 260	+90 = 350			

FIGURE 10-9

Determination of Lags

CPM 2

Cycle	1	2	3	4	5	6	7
Operation to change	10 — 3	13 — 2	12, 13 — 1	12, 15 — 1			
Network limitation	none	1	none	4			
Time changed	3	1	1	1			

Pre-op	Post-op	Pre-EF	Post-ES	Lag	Cyc 1	Lag	Cyc 2	Lag	Cyc 3	Lag	Cyc 4	Lag	Cyc 5	Lag	Cyc 6	Lag	Cyc 7	Lag
1	3	1	1	0														
1	4	1	1	0														
2	7	10	13	3		3		3		3		3						
3	6	7	7	0														
4	5	2	2	0														
5	6	5	7	2		2		2		2		2						
6	7	13	13	0														
6	8	13	13	0														
7	9	15	16	1		1		1		1		1						
8	9	16	16	0														
9	10	21	21	0	1													
10	11	31	31	0	←|←													
10	12	31	31	0	←|←													
11	13	36	36	0	←|←													
12	14	36	36	0	←|←				←|←		←|←							
13	15	40	40	0	←|←		←|←		←|←									
13	17	40	43	3	←|←	3	←|	4	←|←	4	⦶	3						
13	18	40	43	3	←|←	3	←|	4	←|←	4	⦶	3						
14	16	41	41	0	←|←				←|←		←|←							
15	19	48	48	0	←|←		←|←		←|←		←|←							
15	20	48	48	0	←|←		←|←		←|←		←|←							
16	17	43	43	0	←|←				←|←		←|←							
16	18	43	43	0	←|←				←|←		←|←							
17	19	46	48	2	←|←	2	|←	1	←|←	1	←|←	1						
17	20	46	48	2	←|←	2	|←	1	←|←	1	←|←	1						
18	21	46	46	0	←|←				←|←		←|←							
19	22	51	51	0	←|←		←|←		←|←		←-|←							
20	22	50	51	1	←|←	1	←|←	1	←|←	1	←|←	1						
21	22	50	51	1	←|←	1	|⦶	0	←|←		←|←							

FIGURE 10-10

had the least slope of the three, so the other possibility for crashing is restricted to operations 14 and 15 in tandem. The sum of the two slopes represented by the x's opposite operations 14 and 15 is 135. The least expensive of the two possibilities is crashing operation 3 by itself by two days if the network limitation would so allow.

The determination of the network limit as well as the crashing of this small warehouse job through the remaining cycles are demonstrated in Figures 10-11, 10-12, 10-13, and 10-14. It should be noted that at the end of cycle 5, a new zero lag was introduced between the EF of pre-op 5 and the ES of post-op 6, requiring the updating of the circle notation network, Figure 10-13a and b, and the Sequence of Operation worksheet, Figure 10-14.

The process of updating the Sequence of Operation worksheet, form CPM 3, now can be demonstrated in detail. We refer to Figure 10-14, and place a circle in the box of pre-op 5—post-op 6. This graph is updated on account of the new circle in two steps. The first step consists of going down the column of post-op 6 until the diagonal line is met on row 6, proceeding to the right on row 6 and, wherever a circle or an x is encountered, placing an x in the row of pre-op 5 in the column where the circles and x's were encountered. The second step consists of moving to the left on row of pre-op 5 until the diagonal line is met in the column of post-op 5, proceeding up the column until a circle is found in the row of pre-op 4, and moving to the right on the row of pre-op 4 placing x's above all the circles and x's found in the row of pre-op 5.

In the updating of the Operation Selection worksheet, Figure 10-11, the new critical operations 4 and 5 form a new critical subchain of 1, 4, 5, and 6 parallel to critical operations 1, 3, and 6. Operations 1, 4, and 5 cannot be crashed and are struck out immediately.

At the end of cycle 6, a new zero lag is introduced between the EF of pre-op 2 and the ES of post-op 7, necessitating the doubling of the connecting line on Figure 10-13 without creating any new critical subchains. However, this zero lag will not prevent operation 6 from any further crashing, as will be explained a little later in this chapter. For now, the project is crashed through cycle 7, which is as far as it can be crashed with the present information. The project has been crashed a total of 11 days at $925 additional cost. Even though there exist some

Operation Selection Worksheet CPM 4

Critical Paths				Cycle 1	Cycle 2	Cycle 3	Cycle 4	Cycle 5	Cycle 6	Cycle 7
~~1~~		~~1~~								
~~2~~						3	3	③		
		~~4~~								
		~~5~~								
6		6							⑥	
~~8~~										
~~9~~										
~~10~~	~~10~~			⑩						
11										
	~~12~~					⑫	⑫			
~~13~~					⑬	⑬				
	~~14~~							14	14	⑭
15							⑮	15	15	⑮
	~~16~~									
	~~18~~									
19										
	~~21~~									
~~22~~	~~22~~									

Crash-ops

Op.	Slope	Time		Cycle 1	Cycle 2	Cycle 3	Cycle 4	Cycle 5	Cycle 6	Cycle 7
~~12~~	20	~~2~~ 1				ⓧ	ⓧ			
~~10~~	40	~~3~~		ⓧ						
7	60	1								
~~13~~	60	~~2~~ 1			ⓧ	ⓧ				
~~14~~	65	~~2~~						X	X	ⓧ
15	70	~~4 3~~ 1					ⓧ	X	X	ⓧ
17	80	1								
~~3~~	100	~~2~~				X	X	ⓧ		
6	105	~~2~~ 1							ⓧ	
19	110	1								
11	120	2								
	slope			㊵	㊿⁶⁰	100 ⑧⁰	100 ⑨⁰	⑩⁰ 135	⑩⁵ 135	⑬⁵
	time			x3 +	x1 = 4 +	x1 = 5 +	x1 = 6 +	x2 = 8 +	x1 = 9 +	x2 = 11
	add'l cost			120 +	60 = 180+	80 = 260	+90 = 350+	200 = 550+	105 = 655+	270 = 925

FIGURE 10-11

Determination of Lags   CPM 2

Cycle					1		2		3		4		5		6		7	
Operation to change					10 3		13 2		12, 13 1		12, 15 1		3 2		6 2		14, 15 2	
Network limitation					none		1		none		4		2		1		3	
Time changed					3		1		1		1		2		1		2	
Pre-op	Post-op	Pre-E.F.	Post-E.S.	Lag		Lag		Lag		Lag		Lag		Lag		Lag		Lag
1	3	1	1	0														
1	4	1	1	0														
2	7	10	13	3		3		3		3		3	←|	1	|⊕	0		
3	6	7	7	0									←|←					
4	5	2	2	0														
5	6	5	7	2		2		2		2		2	←|⊕	0				
6	7	13	13	0									←|←		←|←			
6	8	13	13	0									←|←		←|←			
7	9	15	16	1		1		1		1		1	←|←	1	←|←	1		1
8	9	16	16	0									←|←		←|←			
9	10	21	21	0									←|←		←|←			
10	11	31	31	0	←|←								←|←		←|←			
10	12	31	31	0	←|←								←|←		←|←			
11	13	36	36	0	←|←								←|←		←|←			
12	14	36	36	0	←|←				←|←		←|←		←|←		←|←			
13	15	40	40	0	←|←		←|←		←|←				←|←		←|←			
13	17	40	43	3	←|←	3	←|	4	←|←	4	|⊕	3	←|←	3	←|←	3	|⊕	1
13	18	40	43	3	←|←	3	←|	4	←|←	4	|⊕	3	←|←	3	←|←	3	|⊕	1
14	16	41	41	0	←|←				←|←		←|←		←|←		←|←		←|←	
15	19	48	48	0	←|←		←|←		←|←		←|←		←|←		←|←		←|←	
15	20	48	48	0	←|←		←|←		←|←		←|←		←|←		←|←		←|←	
16	17	43	43	0	←|←				←|←		←|←		←|←		←|←		←|←	
16	18	43	43	0	←|←				←|←		←|←		←|←		←|←		←|←	
17	19	46	48	2	←|←	2	|←	1	←|←	1	←|←	1	←|←	1	←|←	1	←|←	1
17	20	46	48	2	←|←	2	|←	1	←|←	1	←|←	1	←|←	1	←|←	1	←|←	1
18	21	46	46	0	←|←				←|←		←|←		←|←		←|←		←|←	
19	22	51	51	0	←|←		←|←		←|←		←|←		←|←		←|←		←|←	
20	22	50	51	1	←|←	1	←|←	1	←|←	1	←|←	1	←|←	1	←|←	1	←|←	1
21	22	50	51	1	←|←	1	|⊕	0	←|←		←|←		←|←		←|←		←|←	

FIGURE 10-12

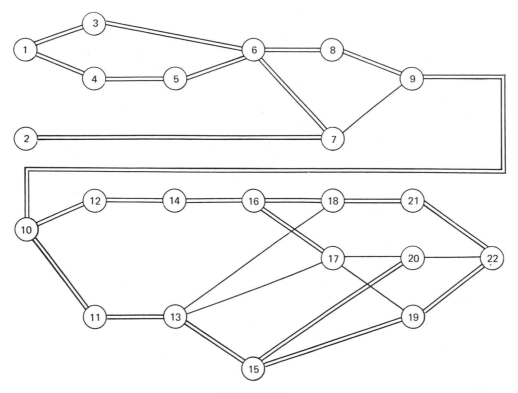

FIGURE 10-13a

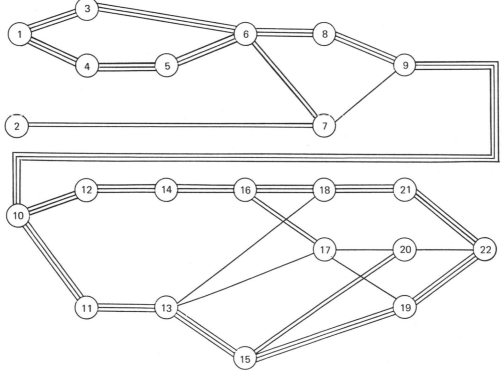

FIGURE 10-13b

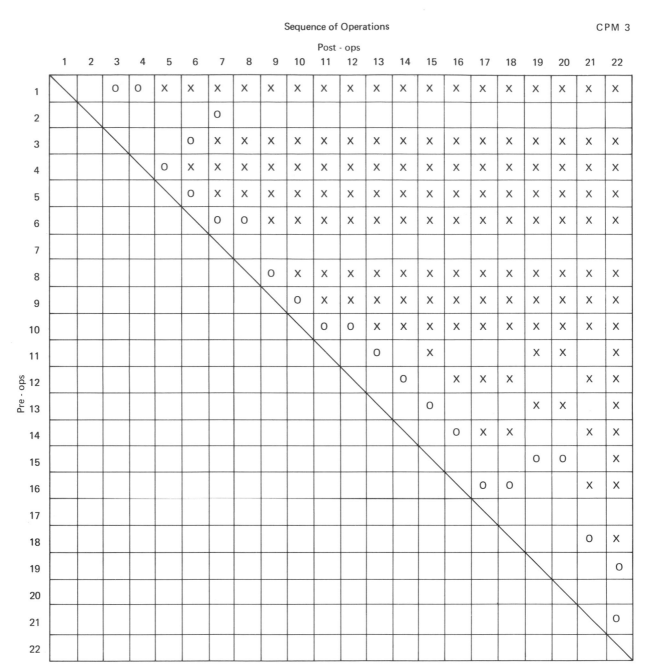

FIGURE 10-14

six lags at the end of cycle 7 and some six operations that can be shortened, it usually would be useless to do so at this stage. Except for
operation 6, either the other operations are noncritical or the effect of
crashing any of the three remaining critical activities would be to make

them noncritical, which would introduce new lags and would not shorten the project duration.

The limit to which a project can be crashed can be determined beforehand by referring to Figure 9-4; use the crash durations to run up the ES and EF of the operations. The limit would be the difference between the earliest finishes of the last operation figured at normal duration and at all crash. In the small warehouse job, the limit is 12 days. The foregoing demonstration shows only 11 days. This is on account of the rare situation of operation 6 with respect to operations 2 and 7. Had operation 2 had a duration of say eight days rather than ten days, the lag between the EF of pre-op 2 and the ES of post-op 7 would have been five days rather than three, allowing operation 6 to be crashed its two days in cycle 6 instead of just one day and thereby giving the expected 12-day project crash without trouble. However, this was not the case.

The circle notation network, Figure 10-13b, shows that operation 6 can yet be crashed. The Determination of Lags worksheet, Figure 10-12, would indicate that the interaction limit of zero does not allow operation 6 to be crashed any further. The fact that the crashing process appears to be completed at 11 days, one day less than that expected from the time and cost comparison in Figure 9-4, signals that something else needs to be done. This is true. If all the worksheets had been reprepared on the basis of the project status, it would be found at the end of cycle 6 that there developed a one-day lag between the EF of pre-op 6 and the ES of post-op 7, necessitating the removal of the doubling line on the circle notation. Also the lag between the EF of pre-op 7 and the ES of post-op 9 would have become zero, requiring doubling of that connecting line. The line connecting operations 2 and 7 already has been doubled as a result of crashing operation 6 by one day. This new sequence of doubled connecting lines is continuous from operations 2 through 7 to 9, creating a new subchain of critical operations in parallel to the two previous chains of operations 1 to 6 and necessitating the further crashing of operation 6 in tandem with operation 7. If all of the worksheets were to be redrafted in the light of the foregoing discussion, it would be possible at the end of cycle 8 to crash this project through to its 12-day limit.

CHAPTER 11

CHANGE ORDERS
AND
ALTERNATES

Very seldom is a building construction project completed in accordance with the original plans and specifications. This is not surprising when one considers all of the elements comprising a turn-key job. Humans unfortunately fail to remember, without some sort of mechanical aid, as few as ten items. It is no wonder then that a set of plans and specifications usually has to be amended after construction is underway in order that errors and omissions in the original documents may be corrected. Many of these errors and omissions are minor and, because the contractor has obligated himself to furnish a complete building "to all intents and purposes" of the contract documents, he will find it good business policy to supply the deficiency at his own expense rather than to take up issue with the architect and the owner. At other times, the error or omission may be of such significance as to warrant a change order in the plan and specifications. If a change in the contract documents is serious enough to warrant a change order, then it is serious enough to warrant a change in the contract price and a revision in the progress schedule. This fact is not always apparent on the face of the change order and the American Institute of Architects' (AIA) standard form of contract recognizes this, so it provides for compensation to the contractor in one of three ways: by a negotiated price, by time and materials, or by unit price. The architect does not have access to the contractor's costs, so it is difficult for the two to arrive at a negotiated price. Compensation by time and materials implies "a blank check" and is, therefore, abhorrent to the owner. The unit price method is the most agreeable to all parties concerned if the unit price is the same as that submitted by the contractor in his proposal. So usually most change orders are written and accepted based on unit prices. However, a change order using the same unit prices contained in the contract is the least profitable to the contractor. It may well lose money for him without his knowing it, unless he revises the critical path schedule to analyze the effect of the change order on apparently unrelated items of work.

For example, in the case of the small warehouse of this book, during the erection of the masonry walls--activity J--the owner decides that he must have a ten-ton hoist installed in the warehouse. The architect immediately prepares plans for the installation of the hoist and finds that some changes need to be made in the masonry walls and in the roof structure--activity K--in order to support the hoist. The contractor probably would negotiate a price for this change order based upon the extra cost of changing the masonry walls and roof structure and the purchase and installation of the hoist. In this instance, he has failed to recognize and account for the effect of this change on the rest of the project. Referring to Figure 7-2 and assuming that the revised plans and the change order are effected prior to completion of masonry walls, we see that there is still a strong possibility that the completion of masonry walls will be delayed and this activity is critical. The following activity, O.W. steel joists, will have its duration lengthened as well as its early start and early finish delayed, and it is also critical. As a matter of fact, the remaining activities in the critical path--the installation of ductwork, carpentry, and interior doors; and painting--all will have their early starts and early finishes delayed, thus extending the project beyond its originally scheduled completion date. The float times on noncritical activities, such as piping and fixtures, wiring and fixtures, furnace and grilles, will be extended and all cause added expense to the contractor, not only in disruption to his subcontractors but also in added overhead due to delay in time on all of the remaining activities.

A change order affecting a noncritical item will have the effect of either increasing its float or reducing its float or even making it critical and changing other activities from critical to noncritical status. For example, should the change order require additional mechanical work that would change the duration of activity Q--piping and fixtures—from three days to six days, activity Q then would become critical and activity O—ductwork—would become noncritical. Activity Q originally had two days of free float and was scheduled to be performed concurrently with activity O, which was critical. Lengthening the duration of activity Q by three days changes the critical path, introducing one day of free float to activity O and lengthening the project duration by one day.

The omission of a noncritical activity would have the effect of increasing only the total float in its subchain without shortening the project duration and, therefore, would not entitle the owner to a reduction in the contractor's overhead cost. The omission of a critical activity would have the effect of making some other operation critical and would not shorten the project duration by the amount of the duration of the omitted activity. It could shorten the duration of the project only by the amount of float in the subchain, which would now become critical. Once again, the change in cost of the project would not be entitled to full credit in an amount proportionate to the duration of the omitted activity.

Another stage in which the critical path method of planning and scheduling can be of help to the builders is the bidding stage. Many owners do not know exactly what they want or can afford. Many architects cannot foretell exactly what the cost of a proposed building will be. The result is that many plans and specifications advertised for bids contain one or more alternates in materials or methods, or both. Yet time is of the essence to the contract. Prospective bidders have not sufficient time to analyze the bidding documents and submit a proposal to the architect by the date set forth for receiving the bids. Yet if time were available to make the study, the bidding contractors would be in a better position to prepare a critical path network for the base bid and each alternate. It is quite possible that the difference in project duration may be of sufficient concern to the owner to warrant or disallow the alternate.

CHAPTER 12

DISTRIBUTION
OF
MEN AND EQUIPMENT

In the course of construction, many things can go awry. A building project may be well underway and proceeding smoothly but the contractor always is bidding on other projects. He successfully bids on a second project and is notified to proceed with the work. He now must coordinate his resources to execute both jobs concurrently. He must plan to shuttle foremen, labor, and equipment back and forth. He would like to plan this shuttling at the least expense to either job and yet complete both on schedule.

A solution to this dilemma is to prepare an arrow network diagram independently for each project and then to draft to time scale the arrow network for each project, as demonstrated in Figure 7-2; but both projects should be drafted on the same chart, bringing calendar-time coordination as well as project activities into focus.

With the time relationship so coordinated between the two projects, the specific requirements of labor and equipment should be entered immediately below each arrow. Now on any given day the resources required for execution of both projects can be summed easily. If the summation of any day's resources is greater than the amount available to the contractor, some additional planning is required. The first thing to look for is float. There always is the possibility that one of the two or more activities requiring the resources has some float time available. If so and if the float is large enough, delaying the start of the activity with the greatest float may be the simplest and most economical solution.

If both the activities requiring the resources are critical and if additional resources are not available, then rescheduling is in order to determine the best and least expensive way out of the dilemma. The interrelationship of activities between the two projects now needs to be indicated by the use of dummy arrows, showing that an activity of one of the jobs cannot start until an activity of the other job is completed. This will require the redrafting of one or both of the schedules, but a CPM schedule, like any other, is useless when it does not consider the exigencies of the builder.

Figure 12-1 applies the foregoing discussion to a practical problem, using just the minimum amount of activities necessary for demonstration. A builder schedules houses number 1 and number 2; the foundation work of house number 2 is to start five days after the first foundation is started. The intent is to start a house every five days. Availability of labor for foundation work is no problem, but the builder has available only six carpenters and helpers with which to execute the work.

At the top of Figure 12-1 is a typical arrow diagram of five activities applicable to any of the houses under consideration. Below this is the same arrow diagram drafted to time scale in working days and identified as house number 1. Below this, the starting five days later, is another time scaled diagram, identical to the first, but identified as house number 2. In both diagrams, the labor requirements are entered below each arrow. At the bottom of the figure is the daily summation of resource requirements. It can be seen that under this plan seven carpenters and helpers will be needed during the 13th and 14th working days. So some changes in scheduling are indicated.

The overrequirement of one carpenter on the 13th and 14th days can be overcome in part by taking advantage of the two days of float in the precutting of rafters and joists. Figure 12-2 shows a solution. The raising of walls, normally a two-day job for three carpenters, is divided into two activities: (1) raising the outside walls which requires two carpenters for one day and (2) raising the interior walls which requires four carpenters working one day. However, these four carpenters cannot start raising interior walls until they have completed framing the

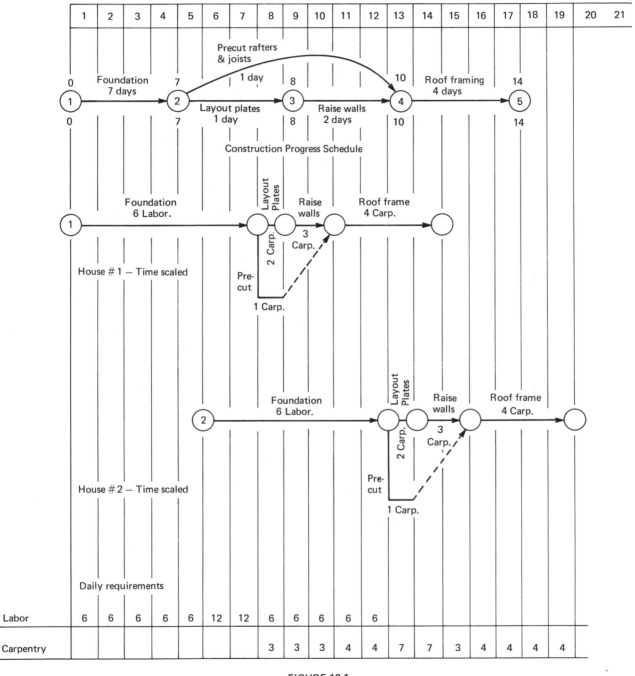

FIGURE 12-1

roof of house number 1 and this restriction is shown by the use of a
dummy activity from node 21 to node 22. Even though this solution may
not utilize the full use of the six carpenters, it does eliminate the overrequirement and still completes house number 2 in the time originally
planned. To this extent, the plan in Figure 12-2 meets the desired
purpose: it accomplishes all of the work in the scheduled time utilizing the
resources available to the builder. Figure 12-2 shows the start of two
houses only. With more houses under construction, it would be possible
to use all six carpenters continuously. This solution is comparatively
simple. But, in reality, the solution to the problem of resource leveling
can become highly complicated. Complications may arise when two or
more concurrent activities require the use of the same type of resource,
in which case management must decide on priority. For example, in
the following illustration

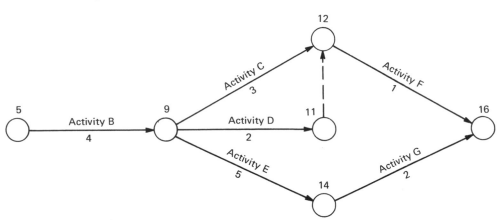

activities C, D, and E may be performed concurrently, yet all three
require the same type of resources over varying durations. The builder
has available sufficient resources to perform any two of the three
activities concurrently. He may decide to replan the schedule, thus:

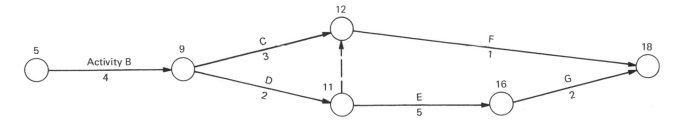

extending the project duration by two days. Or, he has the alternative of replanning, thus:

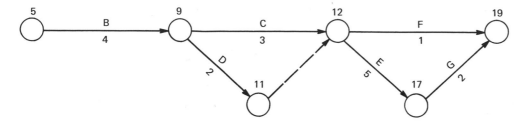

extending the project duration by three days. Or, he has a second alternative of replanning the schedule, thus:

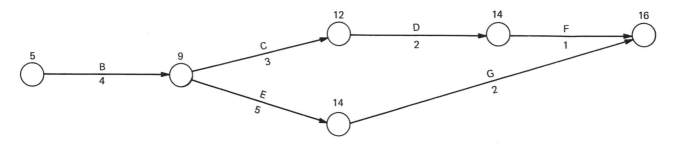

which does not extend the project duration at all because it utilizes float time accruing to activities C and D, and it would be immaterial whether activity C preceded activity D or vice versa. The third solution is the best and is obtained by use of the critical path method of project scheduling.

Another condition that may complicate optimum use of resources could arise when a builder has two or more projects going in different locations. It would be difficult enough for the builder to reschedule his own resources, but when a considerable number of the projects are let to subcontractors who have concurrent contracts with other builders, the solution may not be found without complete cooperation of the subcontractors.

When the problem falls within the builder's own organization, complex problems of resource leveling may be solved by use of a computer. But when the problem involves subcontractors with extraneous interests, the chances of finding a competent solution even with a computer are just as small as those of a participant in a chain letter scheme.

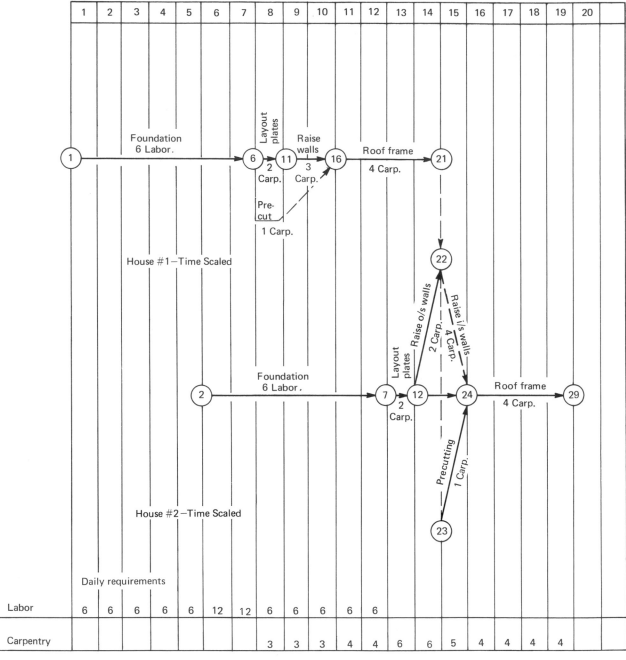

FIGURE 12-2

CHAPTER 13

COMPUTERS
AND
CPM

A computer is a machine. It is a tool that only management can decide when to use or not to use. It is not within the scope of this book to detail the inner workings of the computer or to attempt to make computer experts out of construction men. But because computers have made their way into the industry, it might be helpful to know something about their use.

Basically there are two kinds of computers--analog and digital--but they both perform the same task. The computer performs arithmetic functions with high speed, high accuracy, and high cost. It has found its best use in space navigation, where answers to highly complex problems are required in a matter of seconds. As more and more different types of computers have been developed with more and more varied programs, more and more industries have adopted computers to perform their work with fewer and fewer clerical persons. The transportation industry is now computerized; so is the military. Perhaps the greatest use of computers now is enjoyed by the banking and petroleum industries. Because every computer company has developed a wide variety of programs for accounting and billing, it was inevitable that the construction industry should begin, albeit belatedly, to adopt computer use for accounting purposes. More on that will be discussed later.

The operation of a computer, simply stated, is: The computer must be programmed to perform the specific task at hand. The information is fed into the computer, the computer digests it and then emits the answer. In computer language this is called input and output. Both input and output can be performed in several ways. The two most used are punched cards and magnetic tapes. Recently a new technique called marked sense has been developed for preparing the cards, and it is believed that this new technique will accelerate the use of computers in the construction field. But first, better liaison between computer operators and construction men will perforce need to be secured.

Insofar as management is concerned, CPM is ideally suited for computers and computers are ideally suited for CPM. But computer language is unintelligible to construction men and construction language is meaningless to computer men. When the construction industry sought to utilize the computer to do its work, the computer industry would not or could not understand construction's needs. The notably unsuccessful attempt was made to force the use of computers on the construction industry. After a period of several years, during which construction grew to be the nation's second largest industry second only to defense, construction men began to use the computer for accounting and record purposes. These purposes do not utilize the full time capacity of the computer so the industry, still with some sense of temerity, began to use the computer for CPM.

CPM itself is strange and new to the industry. Perhaps it was not the computer at all that made CPM anathema to construction people. Perhaps it was the fallacious thinking that "What was good for my dad is good enough for me" that prompted construction superintendents to reject CPM. A new breed of construction men is making itself felt in the industry. No longer are the top superintendents carpenters or masons who never finished grammar school. Today, top superintendents are men with college degrees in some area of construction and some of them have had academic courses in computer science and CPM.

Computers are available to almost anyone with a need or desire for their use. They can be owned, leased, shared, or hired. Only giant companies can justify the expense of ownership or leasing. The trend is for a computer company to establish a central computer center with satellite stations in all major cities. The satellite stations receive the input, transmit it to the computer center, and receive the output--all via telephone or teletype. Many universities now have computer centers for teaching computer science as well as for accounting and computerized registration and grading of students. But because of the great speed of the computer the schools cannot use their computers full time so, for a fee, they undertake other business.

To utilize a computer for the critical path method of project planning, scheduling, and control, the builder first must prepare the arrow diagram and assign node numbers and durations. The input data required for processing are the i and j node numbers, the description of the activity, and the duration. This information is transferred to cards, either by hand punching or by marking bubbles on a card with a magnetic pencil to make marked sense cards, so called because sophisticated equipment can scan the cards and do the punching. These marked sense cards and special pencils are quite simple and do not require the use of trained keypunch operators, whom most builders do not have in their employ. The cards are prepared--one for each activity including dummies--and delivered to the computer center or one of its satellites. There, there are many programs already developed from which the builder selects the one that will suit his needs. The builder must be careful here; there are quite a few varying CPM programs and if he cannot find exactly what he wants, he may be asked to make some modifications in order to satisfy an available program. Ordinarily, the output will be delivered to the builder in the form of a printout with information by activities in the following order:

   i j DESCRIPTION DUR   ES   EF   LS   LF   TF   FF

This printout will list the activities in any of several ways: in arithmetic progression order of i numbers, in increasing order of early starts, or in ascending order of late finish. The last printout seems to have merit in that the builder could control the job by late finish of each activity.

Another form of printout uses the input developed from a circle notation network. The input would consist of the operation number, description, normal duration, normal cost, crash duration, and crash cost. The printout comes in two lines, one showing the project schedule, the other operation sequence. This may be used in shortening the project duration because the output can be called for at any desired time of project shortening.

A separate run must be called for at each desired time. This can run into a mountain of printout sheets. A burden is placed on management in this phase of computer use because a computer, though highly sophisticated, is just a machine. It cannot replace construction experience, knowledge, or good judgment.

Unless the computer center is on the builder's premises, the normal time required by a computer center service from receipt of input data cards to delivery of printout sheets is 24 hours. This is so because of card reading, sorting, printing, and waiting, notwithstanding the fact that the computer itself operates just a few minutes. Considering that the work done by a computer in a few seconds requires a few hours by hand, unless the project is of tremendous magnitude or complexity the estimator can perform the arithmetic manually in about the same time and about the same cost as a computer center. A million-dollar project will have about 300 or more activities, and with some practice an experienced estimator can perform all the calculations of the CPM in an eight-hour day. For those reasons, it has been advocated that CPM alone does not justify the use of a computer. However, the decision to use or not to use must be made by management. Some large jobs may best be analyzed manually. Some small jobs are best analyzed by computer. All of the calculations can be performed by computer. The larger the scale of the machine, the faster it can operate and the more options in printout can be provided. However, due to the nature of the memory banks in the computers, it is quite difficult to find a computer large enough to properly perform the crashing process or the resource leveling process. Perhaps the more advantageous use of a computer in scheduling a project is found in programming the computer to print out calendar dates for the early starts and late finishes of each activity—this to serve management. The superintendent could best be served by a graphic schedule, similar to that in Figure 7-2.

Computers also have been found useful on multiunit construction, such as high-rise office buildings or new home subdivisions, where a network can be developed for one floor or one house and used repeatedly—with slight if any modification—to develop a printout for the entire project.

CHAPTER 14

IN SUMMARY—
SOME THOUGHTS
ON CPM

To tell a building contractor that he should learn to use CPM in project planning, scheduling, and control is like telling him that his existing method is no good. It is good. It traditionally has been good. He has become successful using it. CPM and its language of arrows and nodes, restrictions and precedence, and computer input and output are strange words in a strange vocabulary to men in the construction industry.

If you try to explain CPM to a construction man, he will reply: "That is exactly what I have been doing all the time in my head. I don't need any fancy networks or diagrams or curves to run my job. The next rain will knock it all into a cocked hat anyway." And this is true.

On the face of it, it is foolish to tell an old mud slinger that he cannot pour concrete until the forms and reinforcing are in place, or to tell a carpenter that he cannot install rafters until the walls are up. And yet, there is a hunger among these men to learn the secrets of CPM. Because it is strange to them and constructing a building on schedule is so difficult, they look for a panacea but do not find it in CPM. They do not find it in CPM because it is not there. And in their disappointment, they overlook the advantages to them that do lie in CPM.

The most important reason that CPM has not been widely accepted by the construction industry was that it was offered too fast, too soon. Too much was written, too little was inculcated. When the U.S. Department of Defense and the giant corporations started writing into their contracts the requirement that the construction progress schedule be designed to meet the requirements of CPM, contractors were taken by surprise. Frantically, they searched for experts in CPM to prepare this type of schedule. And frantically, anyone who had read articles on CPM in the trade bulletins became an "expert" in CPM. The result was chaotic. I have heard that one large contractor in the early days of CPM spent $20,000 attempting to prepare a CPM schedule for a $5 million job and yet failed to supply a schedule satisfactory to the owner. Usually, in those instances where the contract specifications called for a critical path schedule, the successful contractor submitted any schedule that he thought would satisfy the owner's agent. And once having a CPM schedule approved and accepted, he forgot about it until the job went sour. Not until then did the contractors realize that there was substance to CPM. Now most contractors who have been exposed to it try to prepare their construction progress schedule in accordance with the procedure of CPM to the best of their ability, even without a contract requirement to do so. Even so, they use it mainly as a club over their subcontractors.

One of the reasons that CPM has failed to work in building projects is faulty diagramming. Diagramming can become quite complicated, especially when the logic of CPM is incompletely understood. It seems that the designer of a network is prone to use too many unnecessary dummies when a little more knowledge of construction practices would help him to provide a better diagram. The inexperienced man who relies on the construction superintendent for activity durations (and he needs to) sometimes finds that the superintendent gives him not the best estimate of activity durations but the ones that the superintendent feels will make him look the best in the eyes of the owner.

It is my hope that this book is written in a language all construction men can understand; and in understanding, that they recognize CPM for what it is—not a club but a tool to help plan and carry out a project.

For that reason several variations of CPM have been demonstrated. No one of them will suit everybody. But it is hoped that everybody will find something that will benefit him.

If nothing else, CPM helps the builder to analyze and plan his job before starting construction. The drafting of the arrow network or circle notation does that. Many contracts leave the completion time open and require that the bidder state his own completion time. Too often this completion time is determined by guesswork based on past experience on similar jobs. But completion time is taken into account in comparing competitive bids. Ten days as well as $10 can spell the difference between the award of a contract and the rejection of a proposal.

Time and motion studies have been far too few in the construction industry. Cost accounts have been kept meticulously. A contractor's business is not viable without accurate cost accounts. But it is highly doubtful that these cost accounts are tied in with effort. That is to say, a contractor's cost accounting system is based on his normal efforts and resources and cannot be identified to the inclusion or exclusion of maximum effort. For this reason, crash costs are not as reliable as normal costs. Crash costs can be arrived at only from the product of knowledge and experience. And having arrived at a crash cost, one cannot be too petty or meticulous with it: it is a calculated guess and should be treated as such. But planning and predicting based upon a calculated guess are much better than improvisation. So when a job runs behind schedule, going through the crashing process step by step should give the builder a much better insight into what can be done to bring the project back on time at the least additional cost to him. If time permits, it might be helpful for the builder to plot a curve on graph paper to show the crashing process. Using the horizontal axis for time and the vertical axis for dollars, he could plot each cycle of crashing to develop a time-cost curve from normal to all crash. Such a curve is shown in Figure 14-1. The information for this curve is taken from Figure 10-11. The time-cost curve may prove useful to the builder in determining how far he may want to crash this job, especially if it is compared to a similar curve for overhead and indirect costs or time delay damages.

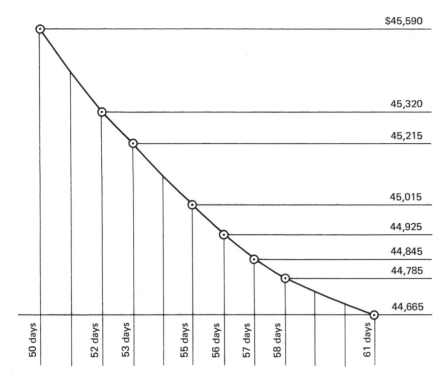

FIGURE 14-1   Time-direct cost curve for a small warehouse.

BIBLIOGRAPHY

1. CPM/A Profitable New Way to Meet Deadlines (Tacoma, Wash.: American Plywood Association, undated).

2. Fondahl, John W. "A Non-Computer Approach to the Critical Path Method for the Construction Industry," Technical Report No. 9 (revised 1962.

3. Hickey, Edgar B. "Critical Path Scheduling: A New Construction Tool," The Navy Civil Engineer (January 1962).

4. Priluck, Herbert M. "CPM Studied, Conceptions and Misconceptions of a New Tool," Building Construction (December 1965).

5. Radcliffe, Byron H., Donald E. Kawal, and Ralph J. Stephenson, Critical Path Method (Chicago: Cahners, 1967).

6. Shaffer, L.R., J.B. Ritter, and W.L. Meyer, The Critical Path Method (New York: McGraw-Hill, 1965).